AF411864

DELTATHERIDIA, CARNIVORA, AND
CONDYLARTHRA (MAMMALIA)
OF THE EARLY EOCENE, PARIS BASIN, FRANCE

DELTATHERIDIA, CARNIVORA, AND CONDYLARTHRA (MAMMALIA) OF THE EARLY EOCENE, PARIS BASIN, FRANCE

BY

THOMAS H. V. RICH

UNIVERSITY OF CALIFORNIA PRESS
BERKELEY · LOS ANGELES · LONDON
1971

University of California Publications in Geological Sciences
Advisory Editors: W. A. Clemens, G. H. Curtis, A. E. J. Engel, R. M. Kleinpell,
J. H. Langenheim, M. A. Murphy, R. L. Shreve, C. A. Wahrhaftig, A. C. Waters

Volume 88

Approved for publication September 17, 1970
Issued March 17, 1971

University of California Press
Berkeley and Los Angeles
California

◇

University of California Press, Ltd.
London, England

ISBN: 0-520-09380-1
Library of Congress Catalog Card No.: 73-633320

CONTENTS

DELTATHERIDIA, CARNIVORA, AND CONDYLARTHRA (MAMMALIA) OF THE EARLY EOCENE, PARIS BASIN, FRANCE

BY

THOMAS H. V. RICH

ABSTRACT

A COLLECTION of previously undescribed fossil mammalian material adds several taxa belonging to diverse higher groups to the known assemblage from the Sparnacian (early Eocene) of France. Unknown before from the Sparnacian but known in the Wasatchian (early Eocene) of North America are the miacids *Viverravus* and *Uintacyon* and the deltatheridian *Didelphodus*. *Hyracolestes* from the (?) early Eocene of Mongolia may have a relative in the Sparnacian.

Two Cuisian (later early Eocene of France) deltatheridians represent new taxa: one is a hyaenodontid, *Francotherium lindgreni*, n. gen. et n. sp., with cheek teeth extremely compressed mediolaterally; the other is an oxyaenid, *Oxyaena menui*, n. sp.

Several taxa previously described from the early Eocene of Europe are represented also in this new sample. Many specimens that can now be assigned only to higher taxonomic categories because the material is inadequate for more refined identification probably represent new genera and species.

SOMMAIRE

UNE COLLECTION de fossiles de Mammifères, non encore décrits, ajoute à l'assemblage connu du Sparnacien (Éocène inférieur) de la France plusieurs unités taxonomiques appartenant à divers groupes plus élevés. Inconnus préalablement du Sparnacien mais connus du Wasatchian (Éocène inférieur) d'Amerique du Nord sont les miacidés, *Viverravus* et *Uintacyon* et le deltatheridien, *Didelphodus*. *Hyracolestes* de l'Éocène inférieur (?) de Mongolie peut avoir eu un ancêtre dans le Sparnacien.

Deux deltatheridiens du Cuisien (Éocène inférieur tardif) sont nouveaux: un hyaenodontidé, *Francotherium lindgreni*, n. gen. et n. sp., avec les dents maxillaires extrêmement comprimées médio-latéralement, et un oxyaenidé, *Oxyaena menui*, n. sp.

Quelques unités taxonomiques de l'Éocène inférieur d'Europe, préalablement décrites, sont representées dans cette nouvelle prise. Beaucoup de pièces qu'on ne peut classer que dans les catégories plus élevées représentent probablement des espèces et des genres nouveaux.

Собрание

Сбор раньше неописанного материала ископаемых млекопитающих увеличивает число родов из разных категорий принадлежащих к известному собранию из Спарнаца (нижний эоцен) Франции. Прежде неизвестные из Спарнаца но известные в Вассаче (нижний зоцен) Северной Америки—Miacidae: *Viverravus* и *Uintacyon*, и Deltatheridia: *Didelphodus*. Может быть у *Hyracolestes* из нижнего эоцена (?) Монголии есть родственник в Спарнаце.

Два Deltatheridia из Квиза (нижний эоцен Франции, позже Спарнаца) представляют новые виды : один—Hyaenodontidae: *Francotherium lindgreni*, n. gen. et n. sp. с поперечно сжатыми ложнокоренными и коренными; другой—Oxyaenidae: *Oxyaena menui*, n. sp.

Некоторые группы, раньше описанные из нижнего эоцена Европы, тоже представлены в этом новом образчике. Многие экземпляры, теперь только вообще отождествленные из-за недостаточного материала, вероятно представляют новые рода и виды.

摘 要

從法國已發現在早始新期 (Eocene) 的化石群中, 另有一些屬於高分類的種屬被收集. 這些是從前還未曾被描述過的哺乳類化石. 其中 Miacidae: *Viverravus* 和 *Uintacyon* 以及 Deltatheridia: *Didelphodus* 還未曾發現在法國的 Sparnacian. 但在北美洲已經知道是在 Wasatchian (早始新期).

蒙古早始新期的 *Hyracolestes* 可能有親屬在法國的 Sparnacian.

兩個 Cuisian (法國後早始新期) Deltatheridia 代表新種類: 一個是有非常扁平的頰齒, 名叫 Hyaenodontid, *Francotherium lindgreni* n. gen. et n. sp. 另一個是 Oxyaenidae, *Oxyaena menui* n. sp.

這個新標本也代表一些從前已經發表過發現在歐洲早始新期的種類. 很多標本可能代表新屬和新種. 它們目前只能被指為高分類的一種, 因為它們已不適於作更細節的檢定.

INTRODUCTION

IN RECENT YEARS paleontologists from the University of California at Berkeley; the Museum National d'Histoire Naturelle of Paris; and Reims, France (Savage, Russell, and Louis, 1965, 1966; Russell, Louis, and Savage, 1967) collected fossil vertebrates from strata in the Epernay district of the Paris Basin. Using the underwater sieving technique described by Hibbard (1949*d*) and McKenna (1962*b*) they were able to recover a large sample of small isolated teeth, as well as several large isolated teeth and mandible fragments with teeth. The material described in this report was obtained largely through their efforts.

Most of the previous knowledge of the deltatheridians, carnivores and phenacodont condylarths from the early Eocene of continental Europe was summarized in three papers by Teilhard de Chardin (1921*a*, 1922*a*, and 1927*a*). The material described by him consisted of less than twenty-five isolated teeth and a single jaw fragment with P_4 and M_1 preserved.

At least three Sparnacian miacids, apparently congeneric with North American Wasatchian forms, are now known: cf. *Viverravus,* cf. *Miacis,* and cf. *Uintacyon.* Among the deltatheridians is a new hyaenodont genus: *Francotherium.* All the other deltatheridian genera have North American affinities: cf. *Didelphodus, Prototomus,* and *Oxyaena. Phenacodus* is another North American form that is present in this assemblage. Most numerous in this collection were the specimens of paroxyclaenids. This exclusively Eurasian family of doubtful ordinal affinities was previously unknown in the Sparnacian. Two forms appear to be present. One has a strong similarity to *Paroxyclaenus* and is represented by four teeth. Fourteen teeth of a second form, which has some similarity to *Spaniella,* are known. Although twice as large, one complete lower molar and a fragment of another display a close morphological similarity to the M_1 of the (?) early Eocene central Asian genus *Hyracolestes.*

Except for one modification by Van Valen (1965*b*, and 1966), I have followed the classification of Simpson (1945*i*). The exclusion of the "creodonts" from the Order Carnivora by Van Valen and his grouping of the palaeoryctids, hyaenodontids, and oxyaenids in the Order Deltatheridia is followed.

The dental terminology proposed by Van Valen (1966:7–9, fig 1, and illustrated here with modifications as fig. 1) was followed throughout this paper. Labial height of the protoconid is the vertical distance from the lowest point on the crista obliqua to the tip of the protoconid (Van Valen, pers. comm., 1967). To describe the measurement from the contact between the root and the crown enamel to the apex of the paracone I have used the term "labial height of the paracone." The various measurements of tooth length and width with one exception were selected from the set of measurements defined by MacIntyre (1966:126–127) for miacids. For the sake of brevity, however, different names were used.

MacIntyre's usage (1966)	This report
Lower molar width (Anterior)	Trigonid width
Lower molar width (Posterior)	Talonid width
Lower molar length	Total length
Upper molar length	Maximum length
Upper molar width (Anterior)	Maximum width
P⁴ length, labial	Maximum length
Premolar length (except P⁴)	Maximum length
Premolar width	Maximum width

Although the definitions of the above terms used in this report are not explicit out of context, no information is lost, for the particular tooth being discussed determines the meaning precisely. The term, "trigonid length," as used in this report refers to the distance from the anterior side of the trigonid in front of the paraconid to the posterior side of the trigonid behind the metaconid.

In order to measure the dip of the anterior cingulum of the various lower cheek teeth for comparative purposes, a standard orientation was assumed for each tooth. Viewing the tooth from directly anterior, the anterointernal corner was assumed to be vertical, and thus a line perpendicular to it along the prevallid was horizontal. The angle of dip was measured relative to such a horizontal line (fig. 2).

All measurements are in millimeters unless otherwise specified.

The material discussed here has been extensively compared with a large number of undescribed and previously described specimens from the early Tertiary of North America and elsewhere. In those cases where comparisons were made with actual specimens, the specimen number is given along with a reference to a

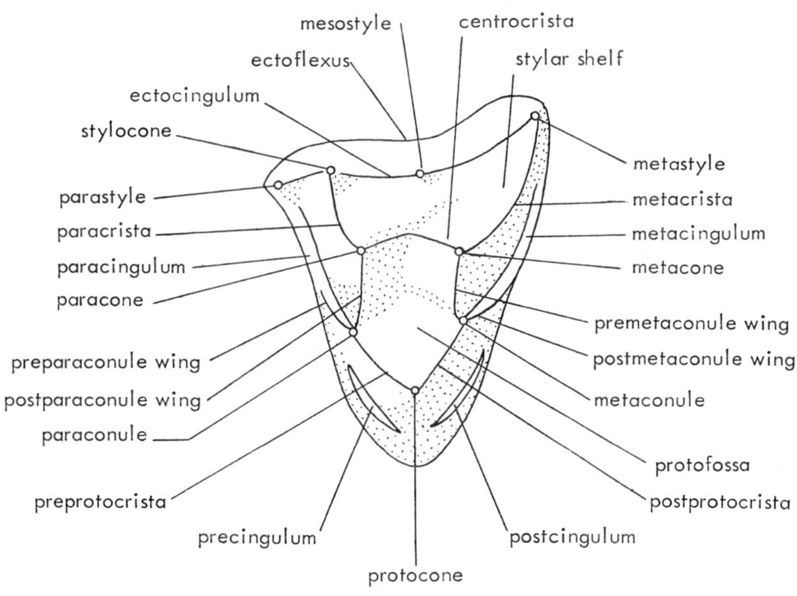

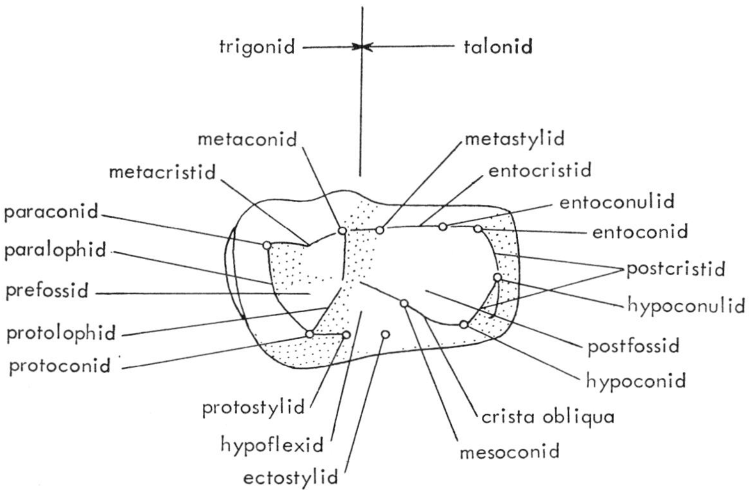

Fig. 1. Diagrams of therian molars showing terminology employed. Trigon is used to designate area encompassed between paracone, metacone, and protocone of an upper molar (top diagram). Anterior side is to the left. Redrawn from Van Valen, 1966, fig. 1.

description or figure if one is known to me. In those cases where only a description or figure of a specimen and not the actual object was compared, only the literature reference is cited.

The following abbreviations are used for institutional collections:

A.M.N.H. The American Museum of Natural History
M.N.H.N. Museum National d'Histoire Naturelle, Paris
U.C.M.P. University of California Museum of Paleontology

ACKNOWLEDGMENTS

I wish to thank the following persons whose assistance has made this monograph possible. Dr. Donald E. Savage suggested the topic and guided its development through every stage with valuable advice and information. Dr. Giles T. MacIntyre spent ten days examining and discussing nearly all the material with me. Drs. Joseph T. Gregory, Seth B. Benson, Frederick S. Szalay, and Robert M. Hunt

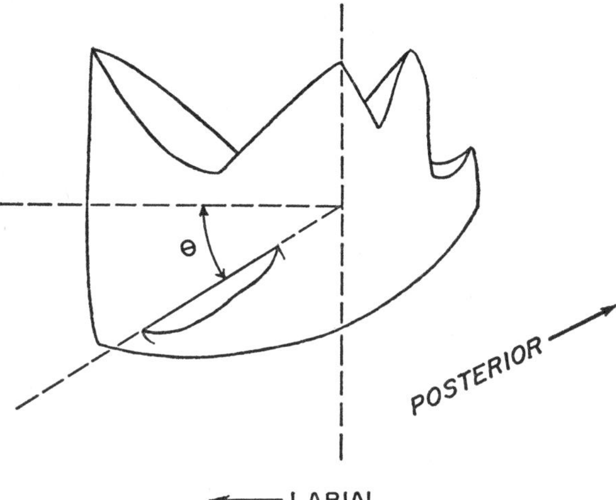

Fig. 2. View of the anterior and lingual sides of a generalized therian lower molar. Θ is the angle of dip of the anterior cingulum.

read this report and made many useful suggestions. Drs. John Dorr and Malcolm C. McKenna supplied needed comparative material. Mr. Owen Poe drew figures 2–18. Mrs. Mary L. Vickers typed several revisions of this report. My wife, Patricia, provided much encouragement throughout the writing and made many valuable comments.

LOCALITIES

The material described in this report comes from seven lower Eocene localities in the Paris Basin. The first six in the list below are in the vicinity of Epernay (Marne) and the last is near Chauny (Aisne).

LOCALITY NAME	LOCALITY NUMBER	STAGE	FORMATION	SITE OF EXTENSIVE UNDERWATER SIEVING
Cuis	......	Cuisan	Sables-à-Térédines	No
Grauves	......	Cuisan	Sables-à-Térédines	No
Mancy	U.C.M.P. V-6255	Cuisan	Sables-à-Térédines	No
Avenay	U.C.M.P. V-6168	Sparnacian	Lignites de Soissonais	Yes
Mutigny	U.C.M.P. V-6167	Sparnacian	Lignites de Soissonais	Yes
Pourcy	......	Sparnacian	Lignites de Soissonais	No
Sinceny	U.C.M.P. V-6962	Sparnacian	Lignites de Soissonais ("sables de Sinceny")	No

TABLE 1

DISTRIBUTION OF TAXA BY LOCALITY

(Numbers followed by asterisk indicate jaw fragments with teeth; numbers without asterisks represent isolated teeth)

	Cuis	Grauves	Mancy	Avenay	Mutigny	Pourcy	Sinceny	Taxa known in North America
Class Mammalia	..	..	..	..	..	..	..	+
Subclass Theria	..	..	..	..	..	..	..	+
Infraclass Eutheria	..	..	..	..	..	..	..	+
Order Deltatheridia	..	..	..	..	..	..	..	+
Suborder Hyaenodonta	..	..	..	..	..	..	..	+
Superfamily Palaeoryctoidea	..	..	..	..	..	..	..	+
Family Palaeoryctidae	..	..	..	..	..	..	..	+
Subfamily Didelphodontinae	..	..	..	..	..	..	..	+
cf. *Didelphodus*	..	..	..	5	5	..	..	+
Superfamily Hyaenodontoidea	..	..	..	..	..	..	..	+
Family Hyaenodontidae	..	..	..	..	..	..	..	+
Subfamily Hyaenodontinae	..	..	..	..	..	..	..	+
Prototomus cf. *P. palaeonictides*	..	1*	..	..	..	..	..	0
cf. *Proviverra* or *Prototomus*	..	..	..	3	2	..	..	+,+
Francotherium lindgreni, n. gen. et n. sp.	..	..	1*	..	..	..	..	0
cf. subfamilies Limnocyoninae or Hyaenodontinae	..	..	..	1	..	..	..	+,+
cf. superfamily Hyaenodontoidea	..	..	..	1	..	..	..	+
Superfamily Oxyaenoidea	..	..	..	..	..	..	..	+
Family Oxyaenidae	..	..	..	..	..	..	..	+
Subfamily Oxyaeninae	..	..	..	..	..	..	..	0
Oxyaena menui, n. sp.	1*	..	..	..	..	..	..	0
Oxyaena sp.	..	..	..	..	..	..	1	+
cf. superfamilies Hyaenodontoidea or Oxyaenoidea	..	1*	..	..	..	..	..	+,+
Order Carnivora	..	..	..	..	..	..	..	+
Suborder Fissipeda	..	..	..	..	..	..	..	+
Superfamily Miacoidea	..	..	..	..	..	..	..	+
Family Miacidae	..	..	..	..	..	..	..	+
Subfamily Viverravinae	..	..	..	..	..	..	..	+
cf. *Viverravus*	..	..	..	1	1	..	..	+
Subfamily Miacinae	..	..	..	..	..	..	..	+
cf. *Miacis*	..	..	..	6	1	..	..	+
cf. *Uintacyon*	..	..	..	2	..	..	..	+
cf. Miacidae	..	..	..	3	1	..	..	+
cf. Deltatheridia or Carnivora	..	..	..	1	..	..	..	+,+
Order Condylarthra	..	..	..	..	..	..	..	+
Family Phenacodontidae	..	..	..	..	..	..	..	+
Phenacodus cf. *P. teilhardi*	..	..	..	..	1*,1	..	..	0
Phenacodus sp.	..	1	..	..	1	..	..	+
Order incertae sedis	..	..	..	..	..	..	..	..
Family Paroxyclaenidae	..	..	..	..	..	..	..	0
cf. *Paroxyclaenus*	..	..	..	2	2	..	..	0
cf. Paroxyclaenidae	..	..	..	6	6	2	..	0
Family incertae sedis	..	..	..	..	..	..	..	..
cf. *Hyracolestes*	..	..	..	1	1	..	..	0

SYSTEMATICS

Class MAMMALIA Linnaeus, 1758*a*
Subclass THERIA Parker and Haswell, 1897
Infraclass EUTHERIA Gill, 1872*b*
Order DELTATHERIDIA Van Valen, 1965*b*
Suborder HYAENODONTA Van Valen, 1967
Superfamily Palaeoryctoidea Van Valen, 1966
Family Palaeoryctidae Simpson, 1931*b*
Subfamily Didelphodontinae Matthew, 1918*h*
cf. *Didelphodus* Cope, 1882*x*

TABLE 2

MEASUREMENTS OF CF. DIDELPHODUS M^1

Specimen	Locality	Maximum length	Maximum width
M.N.H.N.-Av 4917 Right...	Avenay	3.0	4.5

DESCRIPTION

(Fig. 3*a, b*)

M^1—In general outline this tooth resembles an acute angle triangle with the base formed by the labial edge of the tooth. An ectoflexus is developed slightly anterior to the center of the labial margin. The postvallum is subdivided into two straight segments which meet just behind the metaconule to form an oblique angle. Except for where the parastylar spur projects beyond the body of the tooth, the prevallum is straight. Although the paracone is taller than the metacone, the basal dimensions of the two cusps are equal. The paracone cusp is anterior and slightly labial to the metacone. Lowest of the three major cusps, the protocone is slightly less anteriorly placed on the tooth than the paracone. The anterior, posterior, and lingual sides of the protocone pass smoothly into the rounded edges of the adjacent tooth margin. This cusp is located about one-fifth the tooth width from the lingual margin. Rounding occurs only over a small area at the extreme tip of the three major cusps. All three cusps are equally sharp. The preprotocrista curves anterolabially from the protocone to the well-developed paraconule. From there the postparaconule wing curves posterolabially, terminating against the paracone base; and the paracingulum extends labially along the anterior margin of the tooth until meeting the paracrista on the parastylar spur. Extending straight posterolabially from the protocone, the postprotocrista terminates at the metaconule. From there the metacingulum extends in the same direction, terminating on the posterior side of the metacone base. A low centrocrista joins the paracone to the metacone. The stylar shelf is well-developed along the entire labial margin of the tooth. It is widest opposite the metacone. Dividing the paracrista into two segments is a well-developed carnassiform notch at the base of the paracone. The paracrista extends labially from the paracone until

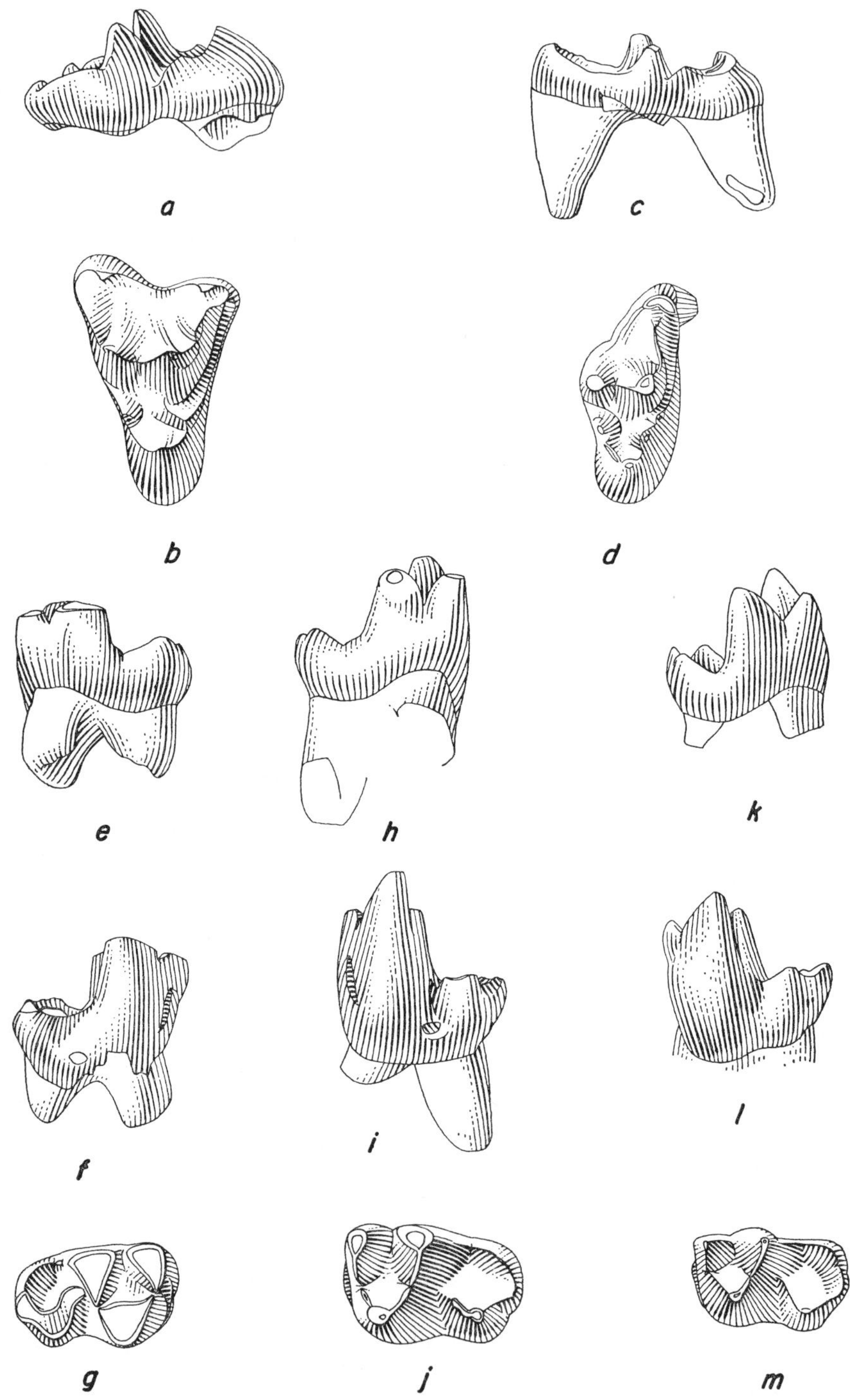

Fig. 3. cf. *Didelphodus*, X 7.5. Right M¹, M.N.H.N.-Av 4917, *a*, posterior view, and *b*, occlusal view. Right M³, M.N.H.N.-Mu 6028, *c*, posterior view, and *d*, occlusal view. Right M₁, M.N.H.N.-Mu 6297, *e*, lingual view, *f*, labial view, and *g*, occlusal view. Left M₂, M.N.H.N.-Mu 6398, *h*, lingual view, *i*, labial view, and *j*, occlusal view. Left M₃, M.N.H.N.-Mu 5502, *k*, lingual view, *l*, labial view, and *m*, occlusal view.

reaching the center of the parastylar spur where it abruptly turns anteriorly, terminating at the paracingulum. If the paracrista had continued labially rather than turning, it would have terminated at the distinct stylocone on the labial margin of the tooth. The metacrista extends straight posterolabially from the metacone almost to the tip of the metastylar spur. Developed along the entire labial margin of the tooth, the ectocingulum is stronger on the flanks of the ectoflexus than at its center. A small mesostyle is developed on the ectocingulum posterolabial to the paracone.

COMMENTS

The M^1 on a specimen of *Didelphodus absarokae* from the Wasatchian of the Big Horn Basin, Wyoming (A.M.N.H. 15700, see Matthew, 1918*h*, fig. 13) is a larger tooth with a greater length relative to its width, connate rather than separate metacone and metaconule bases, postparaconule wing missing, and ectocingulum and metacingulum stronger. In all other ways the two were compared they were alike: relative position, height, and strength of the other cusps and crests; and outline in occlusal view.

TABLE 3

MEASUREMENTS OF CF. DIDELPHODUS M³

Specimen	Locality	Maximum length	Maximum width
M.N.H.N.-Mu 6028 Right..	Mutigny	1.8	4.1

DESCRIPTION

(Fig. 3c, d)

M^3—In occlusal view, the outline of this tooth is an obtuse isosceles triangle with the prevallum forming the base. Both the prevallum and, except for a weak metastylar spur, the posterolabial margin of the tooth are straight. The postvallum is subdivided into two straight segments which meet lingual to the metaconule to form an obtuse angle. The paracone is taller and its basal dimensions greater than those of the metacone. Of the two cusps, the metacone is slightly more lingual on the tooth. Equal to the metacone in height, the protocone is placed lingual to the paracone. There is no sharp break between the adjacent rounded margin of the tooth and the anterior, posterior, and lingual sides of the protocone. Linking the protocone and well-developed paraconule is a short preprotocrista. The prominent paracingulum extends from the paraconule to the tip of the parastylar spur. From the protocone, the postprotocrista curves posterolabially toward the metaconule. From there a short metacingulum extends towards the posterolingual edge of the metacone base. Linking the tips of the paracone and metacone is a low centrocrista. The stylar shelf is broad anteriorly where the parastylar spur is well developed; posteriorly, the shelf is narrow, for no metastylar area is present. A distinct ectocingulum is present along the anterior two-thirds of the stylar shelf. Bases of two or three small stylar cusps are present in the parastylar region.

COMMENTS

Greater size is the most obvious difference between the M^3 of a *Didelphodus absarokae* specimen (A.M.N.H. 15700, see Matthew, 1918*h*, fig. 13) from the Wasatchian of Wyoming and this specimen. Other than that there is only the closer proximity of the metaconule and the metacone bases and the complete absence of any trace of a parastylar spur on the North American specimen to set them apart. Heights, strengths, and relative positions of the various cusps and crests on the two teeth are virtually the same as are their outlines in occlusal view.

TABLE 4

MEASUREMENTS OF CF. DIDELPHODUS LOWER MOLARS

Specimens	Locality	Trigonid width	Trigonid length	Talonid width	Total length
M.N.H.N.-Mu 6297.......... Right M_1	Mutigny	1.9	1.8	1.8	3.0
M.N.H.N.-Av 5872.......... Right lower molar trigonid, M_1?	Avenay	2.0	1.8	...	...
M.N.H.N.-Av 4919.......... Right M_2	Avenay	2.2	1.8	1.8	3.0
M.N.H.N.-Mu 6014.......... Right M_2	Mutigny	2.4	1.9	1.8	3.1
M.N.H.N.-Mu 6398.......... Left M_2	Mutigny	2.1	1.7	1.8	3.0
M.N.H.N.-Mu 5502.......... Left M_3	Mutigny	1.8	1.8	1.6	2.8
M.N.H.N.-Av 5711.......... Right M_3	Avenay	1.8	1.6	1.7	3.0
M.N.H.N.-Av 5870.......... Right M_3	Avenay	1.8	...	1.5	...

DESCRIPTION

(Fig. 3*e–m*)

M_1—The length and width of the trigonid are subequal. The paraconid is smaller in basal dimensions and lower than the metaconid and located anterior and slightly labial to it. The paraconid inclines forward slightly, and its base is somewhat in front of the anterior root of the tooth. The protoconid was probably only slightly taller than the unworn metaconid (as is known to be the case on the M_3) and the two are equal in basal dimensions; the protoconid occupies the labial third of the trigonid and is anterolabial to the metaconid. The prefossid was well developed with the paralophid, protolophid, and metacristid forming the anterior, posterior, and lingual walls respectively. Weak notches are variably present on the metacristid, paralophid, and protolophid, no one tooth having notches on all three crests. There is a faint appression fossette on the lingual half of the anterior face of the trigonid and a weak to moderately developed anterior cingulum on the labial half that dips downward laterally at an angle from 25 to 60 degrees below the horizontal.

The talonid is wider than long, the ratio of these dimensions being about 3:2, its width about one-half to two-thirds the length of the trigonid. A notch is present on the entocristid of one specimen (M.N.H.N.-Mu 6297). Forming a continuous ridge along the lingual margin of the talonid, the entocristid and postcristid extend from the posterior edge of the metaconid to the midpoint of the posterior border of the tooth, the entoconid and hypoconulid being slightly to well-differentiated cusps on this crest. Labial to the midpoint of the posterior border of the tooth, the postcristid is directed anterolabially towards the hypoconid, which is near the lateral edge of the talonid. This labial part of the postcristid is noticeably lower than the lingual on worn teeth. The crista obliqua is directed anterolingually from the hypoconid and abuts against the posterior side of the trigonid near the base midway between the labial and lingual edges of the tooth. The postfossid is deeply enclosed by the surrounding crests and restricted to the lingual two-thirds of the tooth.

M_2—This tooth is a duplicate of M_1 except that the trigonid is significantly wider than long.

M_3—Compared with the M_1, this tooth has a more posteriorly displaced hypoconulid and the talonid is narrower relative to the trigonid.

COMMENTS

These specimens are quite similar in size and morphology to the lower molars of *Didelphodus altidens* from the Wasatchian and Bridgerian of Wyoming. The only apparent difference is that on the Wyoming specimens (A.M.N.H. 12091, see Matthew, 1909*d*, pl. 49, fig. 1; A.M.N.H. 14747, see Matthew, 1918*h*, fig. 15; A.M.N.H. 56614, 56615) the M_2 and M_3 talonid widths are less than their respective trigonid widths whereas these dimensions are more nearly equal on the French specimens. This slight difference may be indicative of an evolutionary trend within a single species toward or away from a narrower talonid but the few specimens known cannot confidently establish the existence of such a trend.

The larger size and more lingual position of the paraconid relative to the metaconid distinguishes these lower molars from those of pantolestids.

Superfamily Hyaenodontoidea Leidy, 1869*a*
Family Hyaenodontidae Leidy, 1869*a*
Subfamily Hyaenodontinae Trouessart, 1885*a*
Prototomus Cope, 1874*o*
Prototomus cf. *P. palaeonictides* (Lemoine, 1880*a*)

TABLE 5

MEASUREMENTS OF PROTOTOMUS CF. P. PALAEONICTIDES

Specimen	Locality
M.N.H.N.-Louis-195 Gr Right Mandible	Grauves

Alveolar measurements

	Width of anterior alveolus	Length of anterior alveolus	Width of posterior alveolus	Total length
P_1............	0.6	1.6	1.2	3.7
P_2............	1.5	1.7	1.6	4.2
P_4............	1.9	2.2	1.8	5.1
M_1............	2.0	1.5	...	Distance from posterior edge of rear root to anterior side of front alveolus $5.0\pm.3$

Tooth measurements

	Maximum width		Maximum length	
P_3..........................	2.1		5.6	
	Trigonid width	Trigonid length	Talonid width	Total length
M_2..........................	3.5	2.9	3.2	$5.6\pm.3$
M_3..........................	3.4	$2.8\pm.3$	2.6	$5.9\pm.3$

DESCRIPTION

(Fig. 4*a–c*)

The specimen (M.N.H.N.-Louis-195 Gr) is a right mandible from Grauves complete from the broken alveolus of the canine nearly to the angle and lacking the ascending ramus above the height of the tooth row. The P_3 is present together with the slightly damaged M_2 and M_3. The crowns of the remaining teeth are missing. The posterior root of the M_1 with a fragment of the posterior part of the talonid projects above the alveolus; the rest of the teeth posterior to the canine are represented only by complete alveoli.

Between the M_2 and M_3 the mandible is 7.6 mm deep and has a maximum width of 4.7 mm. It gradually becomes shallower and thinner anteriorly until it is only 5.7 mm deep and 4.1 mm wide beneath the P_1. A mental foramen is present below the anterior root of P_2 about halfway to the ventral border of the mandible. Below the midpoint of the P_3 and halfway to the ventral border is a second mental foramen. The masseteric fossa covers all the preserved lateral surface of the mandible posterior to a point 4 mm behind the M_3 and dorsal to a line 3 mm above the ventral border. The posterior margin of the symphysis is not clearly marked but occurs near a point below the posterior edge of the P_2. From there the symphysis extends anteriorly over the lower two-thirds of the lingual side of the mandible.

The mandibular foramen occurs at the level of the alveolar row, about 10 mm behind the M_3.

P_3—A small cusp is on the anterior end of this tooth and a short, transversely trenchant talonid lies behind the high main cusp.

M_2—This is the best preserved molar. Its trigonid is slightly wider than long. Comparison of paraconid and metaconid heights cannot be made because the tip of the latter cusp is missing. The basal dimensions of the two cusps are equal. The paraconid is slightly labial to a point anterior to the metaconid. Its base projects slightly forward from the anterior root. The tip of the protoconid is broken, preventing height comparisons with other cusps. Its basal dimensions are somewhat greater than those of the metaconid. The protoconid is one-fourth the trigonid width lingual to the labial border of the tooth, labial and slightly anterior to the metaconid. The apex of the paraconid shows signs of wear but was probably sharp before abrasion. The prefossid is open anteriorly, posteriorly, and lingually. Extending posterolabially from the paraconid is a well-developed paralophid, but it is not met by a similar lophid from the protoconid; therefore the anterior opening of the prefossid is much wider than a carnassial notch. No sign of a protolophid is present between the protoconid and metaconid, but a small one developed on the metaconid would have been obliterated by the forces which damaged that major cusp. Two isolated molar trigonids described below that may belong to the same species as this mandible have prominent carnassial notches on both the well-developed paralophid and the equally strong protolophid (see fig. 5c–e). Near the base of the crown on the lingual side of the prevallid is an appression fossette; immediately labial to which is the anterior cingulum that dips downward laterally at 55 degrees to the horizontal and extends nearly to the anteroexternal corner of the tooth.

The talonid is slightly wider than long, and its length and width respectively are somewhat less than those of the trigonid. The crista obliqua extends from the hypoconid straight to the posterior face of the trigonid abutting below a point about two-thirds the distance from the metaconid to the protoconid. The hypoconid, like the crista obliqua, is heavily worn but it was the talonid cusp with the greatest basal dimensions. When unworn, the height of the hypoconid was probably about equal or slightly less than that of the other two talonid cusps. The hypoconid is located about one-fourth the talonid width lingual to the labial edge of the tooth. The hypoconulid is slightly lower and subequal in basal dimensions to the entoconid. It is located on the posterior border of the tooth slightly lingual to the centerline of the talonid. The entoconid is adjacent to it and in the posterointernal corner of the tooth. Extending anteriorly from the entoconid is a short entocristid. The postfossid is shallow, large, and open on the lingual side of the tooth behind the trigonid. There is no trace of an ectostylid on this tooth. Wear surfaces are prominent on the anterior and posterior side of the trigonid and on the dorsal and anterior surfaces of the hypoconid and crista obliqua.

M_3—This tooth differs from the M_2 primarily in having a markedly narrower talonid. There may have been differences in trigonid morphology that have been obscured by the extensive damage done to that structure on the M_3.

Comparison of the alveoli, posterior roots, and posterior portions of the talonids

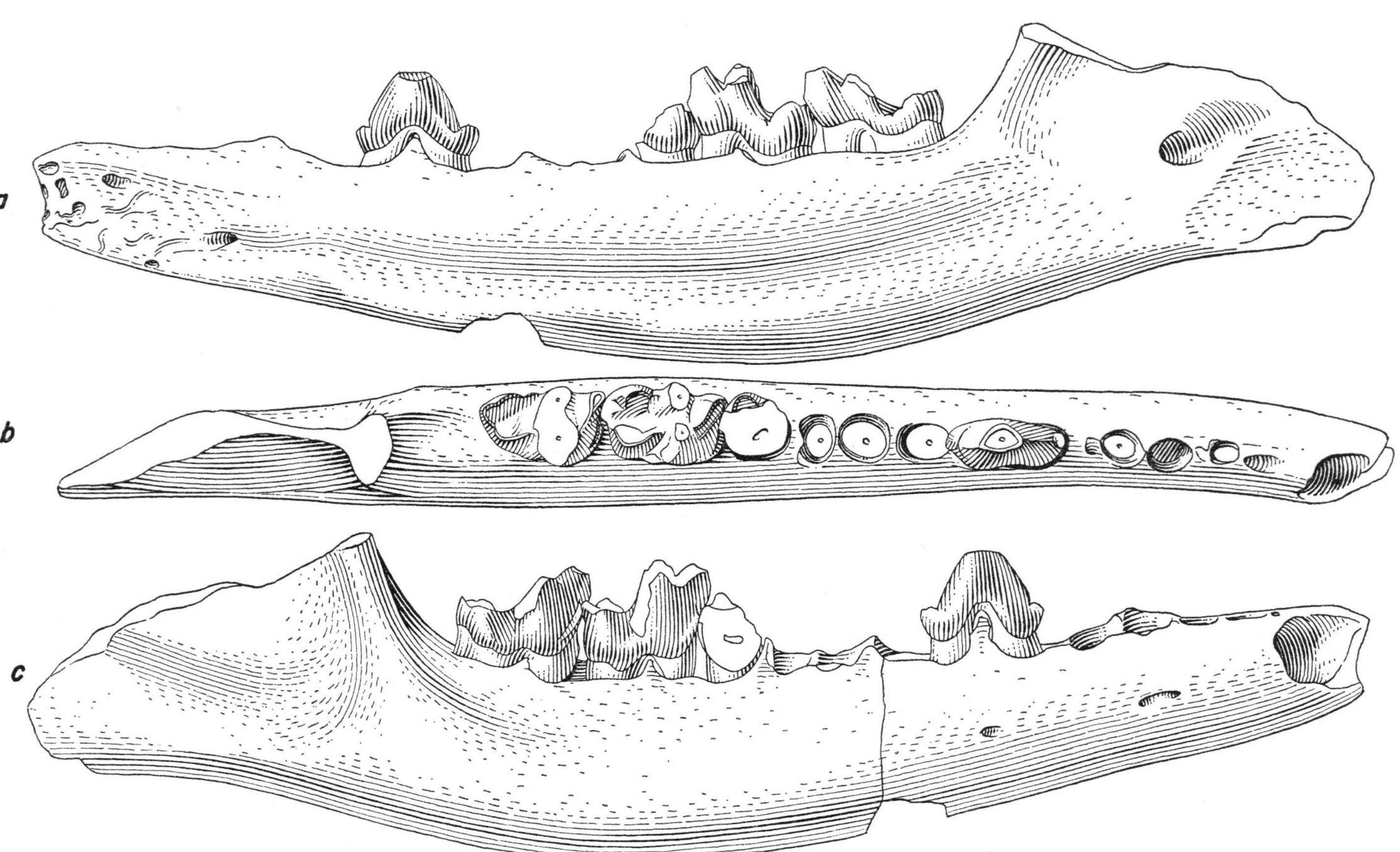

Fig. 4. *Prototomus* cf. *P. palaeonictides*, X 3, right mandible, M.N.H.N.-Louis-195 Gr, *a*, lingual view, *b*, occlusal view, and *c*, labial view.

of the M_1 and M_2 show that these structures are similar in size and morphology, and therefore the M_1 was presumably very similar to the M_2. The alveoli of P_4 are narrower relative to their length than those of the M_1, more like those of P_3. Such a difference indicates that the tooth was probably premolariform. The presence of P_3 makes direct comparison of the alveolar dimensions of this tooth with those of the P_4 difficult, but allowing for an error of ±0.2 mm, they are equal in length and width; presumably the teeth were about the same size. The P_2 is separated from the P_3 by a small diastema; the alveoli of the P_2 indicate that it was a slightly narrower tooth than the P_4 and about four-fifths its length. The P_1 was smaller still, its alveoli indicating a length slightly less than four-fifths and a width two-thirds that of the P_4. The preserved alveolus of the canine is elliptical in cross section with the major axis directed dorsoventrally and measuring 3.2 mm; the minor axis trends in the mediolateral direction and is about 1.9 mm long.

COMMENTS

Van Valen (1965b:642) could find no character to distinguish the lower dentitions of the genera *Prototomus* and *Proviverra*. Therefore, the species *Prototomus palaeonictides* may in fact be a member of the genus *Proviverra*. However, I have provisionally followed Van Valen (1965b:639) in referring *P. palaeonictides* to the genus *Prototomus*.

The type specimen of *Prototomus paleonictides* from the early Eocene of France consists of a mandible fragment with the P_4 and M_1 preserved. Therefore, the Grauves mandible, which lacks these teeth, can at best be tentatively referred to that species. In contrast to the M_2 of the Grauves mandible, the M_1 of *P. paleonictides* type (U.C.M.P. 49725, cast) has a lower paraconid relative to the length of the trigonid, a smaller angle between the postvallid and the horizontal plane of the tooth, and a greater angle between the prevallid and postvallid. Similar degrees of difference in all three characters can be found between the M_1 and M_2 of an individual hyaenodont (cf. *Prototomus multicuspis*, A.M.N.H. 16820, see Matthew, 1915d, fig. 73; *?Prototomus* cf. *P. secondaria*, A.M.N.H. 16120). The two teeth are similar in talonid construction. The type specimen of *P. paleonictides* is larger, for the P_4 to M_2 length is 18 mm (*fide* Teilhard de Chardin, 1922a:50) in contrast to 15 mm on the Grauves specimen.

TABLE 6

MEASUREMENTS OF CF. PROVIVERRA OR PROTOTOMUS M^2

Specimen	Locality	Maximum length	Maximum width
M.N.H.N.-Av 4631 Left....	Avenay	6.7	...

DESCRIPTION

(Fig. 5a, b)

The region of this M^2 lingual to the paracone and metacone bases is missing. Also missing are the marginal tips of the parastylar and metastylar regions. The part of the labial border remaining has a deep ectoflexus and the parastylar and

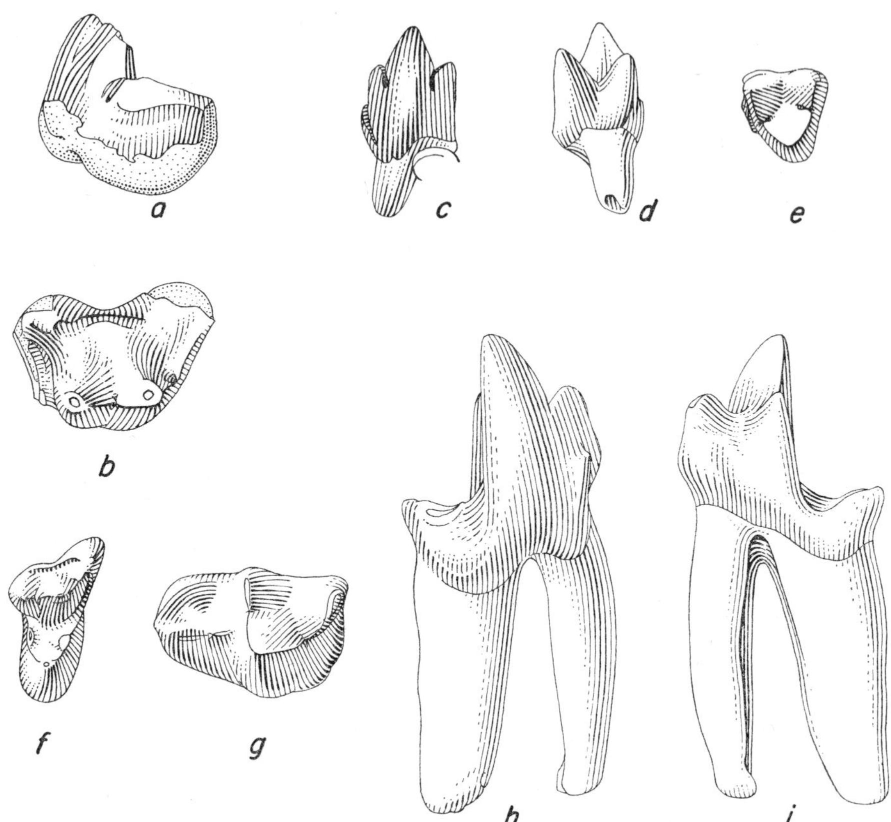

Fig. 5. cf. *Proviverra* or *Prototomus*. Left M², M.N.H.N.-Av 4631, X 4, *a*, posterior view, and *b*, occlusal view. Left molar trigonid, M.N.H.N.-Av 5084, X 4, *c*, labial view, *d*, lingual view, and *e*, occlusal view. Right M³, M.N.H.N.-Mu 6464, X 5.33, *f*, occlusal view. Right lower molar, M.N.H.N.-Av 5070, X 4, *g*, occlusal view, *h*, labial view, and *i*, lingual view.

metastylar spurs extend outward from the main body of the tooth. The paracone is anterior to and taller than the metacone. Abrasion has flattened the tips of both cusps slightly. Developed along the preserved anterior border of the tooth, the paracingulum becomes markedly narrower labially and probably did not reach the marginal tip of the parastylar spur. Uniting with the ectocingulum in the corner of the parastylar spur, the paracrista extends from there to the anterior side of the paracone. Along the posterior border of the tooth, there is only a faint indication of the metacingulum. Separating the posterolabial side of the metacone from the metacrista is a deep carnassial notch. The metacrista extends to the preserved tip of the metastylar region. The ectocingulum is present along the margin of the broad stylar shelf from the tip of the parastylar spur to a point labial to the metacone. A cuspule is developed on the ectocingulum near the point where that crest joins the paracrista. A small short ridge distinct from the ectocingulum is directed posterolingually from this cuspule. A well-developed centrocrista bisected by a notch is present between the paracone and metacone.

COMMENTS

The prominent parastylar spur, the relatively low angle between the metacrista and a line from the paracone to the metacone, the deep ectoflexus in the labial border, and the connate paracone and metacone are characters in common between this specimen from Avenay and an M^2 of *Prototomus multicuspis* (A.M.N.H. 15239, see Matthew, 1915*d*, fig. 72). The two specimens differ primarily in that the North American one lacks a paracrista.

TABLE 7
MEASUREMENTS OF CF. PROVIVERRA OR PROTOTOMUS M³

Specimen	Locality	Maximum length	Maximum width
M.N.H.N.-Mu 6464 Right..	Mutigny	2.0	4.4
M.N.H.N.-Mu 6368 Left....	Mutigny	1.8	...

DESCRIPTION
(Fig. 5f)

In occlusal view the outline of this M^3 is an obtuse triangle with the base formed by the prevallum. The parastylar region projects far beyond the main body of the tooth, the labial side is emarginated by an ectoflexus, the base of the metacone extends far to the rear of the adjacent posterior margin of the tooth, and the anterior edge is straight. The paracone is much higher and larger in basal dimensions than the metacone. Its labial height is three-fourths the tooth length, and it is somewhat labial to a point directly anterior to the metacone. The tips of the paracone and metacone are rounded. The protocone is lower and subequal in basal dimensions to the metacone. The precingulum and postcingulum are weakly developed. Linking the large paraconule to the protocone is a short preprotocrista. Uniting with the paracrista at the point two-thirds the distance from the paracone to the parastyle, the wide paracingulum extends along the anterior margin of the tooth from there to the paraconule. The postprotocrista extends posteriorly from the protocone to the posterior margin of the tooth and there turns labially and nearly joins the metacone from which it is separated by a prominent, narrow, deep notch. Extending from the paracone to the parastyle, the paracrista has two slight notches developed in it. From the parastyle, the ectocingulum extends posteriorly along the outer margin of the tooth for about two-thirds its length. No trace of a metacrista, metastyle, or ectocingulum is present in the metastylar region. With the exception of a prominent expansion in the parastylar area, the stylar shelf is rather narrow. The centrocrista is present but low.

COMMENTS

The M^3 of miacids is not as transverse as these specimens nor does it have as prominent a parastylar spur. Except for these characters there is no reason to place these specimens in the hyaenodontids rather than the miacids. They more closely resemble the M³ of *Prototomus multicuspis* from the Wasatchian of Wyoming (A.M.N.H. 14773, see Matthew, 1915*d*, fig. 72) than the M³ of any other species of hyaenodontid examined. The only differences are that the M³ of *P. multicuspis* is twice the size of these teeth and has a straight labial margin with no indication of an ectoflexus.

TABLE 8

MEASUREMENTS OF CF. PROVIVERRA OR PROTOTOMUS LOWER MOLAR

Specimen	Locality	Trigonid width	Trigonid length	Talonid width	Total length
M.N.H.N.-Av 5070 Right..............	Avenay	4.1	4.0	3.6	6.5

DESCRIPTION

(Fig. 5*g–i*)

The width and length of the trigonid are the same. The paraconid is the lowest trigonid cusp and is slightly larger in basal dimensions than the metaconid. It is located anterior and a trifle lateral to the metaconid. The cusp points vertically, and its base is slightly forward of the anterior root. The protoconid is the highest and largest trigonid cusp, its labial height above the talonid being one and one-third the trigonid length. It is situated one-fourth the trigonid width medial to the labial edge of the tooth and anterolabial to the metaconid. The protoconid has a sharp tip, the paraconid has a rounded one, and the metaconid is intermediate in shape. The crests of the paralophid and protolophid dip only slightly downward from the paraconid and metaconid respectively to small carnassiform notches. However, the crests of these same lophids dip steeply downward from the protoconid to the carnassiform notches. The prefossid is open lingually owing to a prominent gap between the paraconid and metaconid. A prominent anterior cingulum dipping downward labially at an angle of 60 degrees below the horizontal is developed on the lateral half of the prevallid.

The talonid is two-thirds as long as it is wide. It is three-fifths as long and nearly as wide as the trigonid. The entocristid and postcristid form a low, unbroken arc along the lingual and posterior edges of the talonid from the base of the metaconid to the hypoconulid. From the hypoconulid the postcristid extends anterolaterally to a small mammelon which may be homologous to the hypoconid. Extending directly anterior from this mammelon, the crista obliqua abuts against the posterior face of the trigonid at a point slightly lateral to the carnassiform notch and midway between the metaconid and protoconid. The hypoconulid is the only cusp definitely developed on the talonid. It is in the middle of the posterior border of the tooth. The postfossid is shallow but completely enclosed and is restricted to the lingual half of the talonid. No sign of an ectostylid or wear facet is to be seen.

COMMENTS

This specimen agrees completely with the diagnosis of deltatheridian lower molars given by Van Valen (1966:94–95). The specimen shows a strong resemblance to the lower molar from the early Eocene of Belgium questionably referred to *Sinopa* by Teilhard de Chardin (1927*a*:21, fig. 18*c*). The two specimens are similar in length, the labial height of the protoconid above the talonid, the length ratio of the trigonid and talonid, and the character of the postfossid. In both specimens the anterior root is markedly narrower anteroposteriorly than the posterior one and slightly shorter. The only difference between these two specimens is that the paraconid of the French tooth is a trifle lower than the metaconid while the reverse is true on the Belgian one. The fact that the talonid is only half the length of the trigonid suggests that these two specimens are not referable to *Proviverra* (=*Sinopa* in part) but to a more advanced proviverrine. The low entocristid and postcristid together with the small angle between the prevallid and postvallid immediately distinguishes the French tooth from all described proviverrines (Van Valen, 1965*b*:641–644).

To a lesser degree, this specimen resembles the M_1 of the miacid *Vassacyon promicrodon* (A.M.N.H. 15161, see Matthew, 1915*d*, fig. 36) from the Wasatchian of Wyoming. They are similar in the heights, basal dimensions, and disposition of the trigonid cusps; the development of the protolophid; and the prominence and position of the anterior cingulum. It differs from the M_1 of *V. promicrodon* in the weakness or absence of all three talonid cusps; smaller size, about three-fourths that of the M_1; slightly greater labial height of the protoconid above the talonid relative to the length of the trigonid; and slightly narrower trigonid width relative to total length.

TABLE 9

MEASUREMENTS OF CF. PROVIVERRA OR PROTOTOMUS LOWER MOLAR TRIGONIDS

Specimens	Locality	Trigonid width	Trigonid length
M.N.H.N.-Av 5084 Left................	Avenay	3.1	2.7
M.N.H.N.-Av 5911 Left...............	Avenay	3.4	2.9

DESCRIPTION

(Fig. 5*c–e*)

The length of the trigonid is somewhat less than the width. The paraconid is slightly lower than the metaconid. The two cusps have equal basal dimensions and are well separated from one another. The base of the paraconid projects anteriorly from the mesial root of the tooth. Higher and larger than either of the other trigonid cusps, the labial height of the protoconid above the talonid is one and one-half times the trigonid length. The apices of the trigonid cusps are dull points. The tip of the protoconid is located labial and slightly anterior to the metaconid and is situated about one-fourth the trigonid width medial to the labial edge of the tooth. The prefossid is moderately deep with a well-developed paralophid and protolophid (both of which have a prominent carnassiform notch) forming the anterior and posterior walls respectively. The prefossid is open on the lingual side of the tooth between the widely spaced metaconid and paraconid. The anterior cingulum is developed on the lateral two-thirds of the anterior face of the trigonid near the base and dips laterally 35 degrees below the horizontal.

COMMENTS

The features comparable on these trigonids and that of the M_2 of the jaw referred above to *Prototomus* cf. *P. palaeonictides* suggest the assignment of all three specimens to the same species. They are similar in size, development of the anterior cingulum, height and shape of the paraconid, and anterior projection of the paraconid base from the mesial root.

Francotherium, new genus

Etymology.—Franco, med. L., France, referring to the country of origin; Θηριον, G., beast.
Type.—*Francotherium lindgreni,* new species

Diagnosis.—Differs from other hyaenodontids in the subequal length of the three lower molars; absence of a metaconid on the M_2 and M_3; and the extremely narrow, bladelike aspect of the most posterior two molars.

Francotherium lindgreni, new species

Etymology.—Named in honor of Dr. Frank T. Lindgren.
Type.—M.N.H.N.-Louis-49 Ma, a right mandible complete from the posterior side of the canine alveolus to the mandibular foramen with a complete M_3, a nearly complete M_2 and P_4, and the alveoli of P_1–P_3 and M_1.
Type locality.—Mancy, Sables-a-Térédines, Cuisian (latest early Eocene), Paris Basin.
Referred specimens.—None.

Diagnosis.—Only known species of genus.

TABLE 10

DENTAL MEASUREMENTS OF FRANCOTHERIUM LINDGRENI

Specimen	Locality
M.N.H.N.-Louis-49 Ma Right mandible	Mancy

Alveolar measurements

	Width of anterior alveolus	Length of anterior alveolus	Width of posterior alveolus	Total length
P_1	2.0±.3	2.9±.3	2.0±.3	8.1±.3
P_2	2.1	3.3	...	...
M_1	4.4	3.5	...	10.4±.3

Tooth measurements

	Trigonid width	Trigonid length	Talonid width	Total length
M_2	4.5	8.0	3.4	11.8
M_3	4.9	8.6	2.8	10.5

	Maximum width		Maximum length	
P_4	3.2		11.2	

DESCRIPTION

(Fig. 6a–c)

Between M_2 and M_3 the mandible is 22 mm deep and has a maximum width of 9 mm; moving forward, the width and depth remain unchanged until the P_3 is reached where the jaw becomes slightly shallower. There is one large mental foramen directly below the posterior root of the P_1 about halfway to the ventral border of the jaw. Due to missing portions of the mandible below P_2–P_4, the presence or absence of a second mental foramen opening could not be determined. The masseteric fossa occupies the entire lateral surface of the mandible posterior to a point about 8 mm behind the M_3 except for the ventral-most 8 mm. As is characteristic of other European hyaenodonts, the left and right mandibular rami were not fused at the midline symphysis. The symphysis begins below the posterior root of the P_2 and occupies all the lingual surface of the mandible anterior to that point except for a narrow strip about 4 mm wide next to the alveolar row. The mandibular foramen is low, 11 mm above the ventral border of the jaw and about 19 mm behind the M_3. There is a small cusp at the anterior end of the P_4 and a short, trenchant talonid behind the high main cusp. The tooth is quite narrow and not at all molariform.

M_3—This tooth is the only complete molar. Its trigonid is nearly twice as long as wide. The paraconid is anterior and slightly lingual to the protoconid. It is smaller in basal dimensions and much lower than the protoconid and barely higher than the lowest point on the paralophid. The base of the paraconid extends forward of the anterior root of the tooth. There is no metaconid or prefossid on this blade-like tooth. A very slight swelling is present on the prevallid immediately lateral to the point where the talonid of the M_2 appresses against it.

The talonid is three-fourths as long as it is wide. It is only three-fifths as wide and one-fourth as long as the trigonid. This miniscule talonid, a mere swelling on the posterior side of the trigonid, has a single distinct cusp near the middle of the posterior edge of the tooth. There are no distinct indications of wear.

M_2—The M_2 differs from the M_3 in the following details: the trigonid is not quite so long relative to its width or to the total length of the tooth; a small cusp, barely discernible on the M_3, is clearly developed on the anterior side of the paraconid base; the talonid is trenchant and much more prominent; and the tips of the paraconid and protoconid are heavily worn.

The distance from the forward edge of the anterior alveolus to the rear margin of the posterior alveolus of the M_1 is subequal to similar measurements taken on the roots of M_2. The space allowed for the M_1 between the preserved adjacent teeth is subequal to the M_2 in length. The anterior alveolus of M_1 is 0.7 mm narrower than that of M_2. These comparisons indicate that the M_1 was a tooth subequal in length to the M_2 with a somewhat narrower trigonid.

COMMENTS

The subequal length of the M_1–M_3, the absence of a metaconid on M_2 and M_3, and the high tooth length to trigonid width ratio of the M_2 and M_3 separates this specimen from the previously described genera of hyaenodontids. *Propterodon irdinen-*

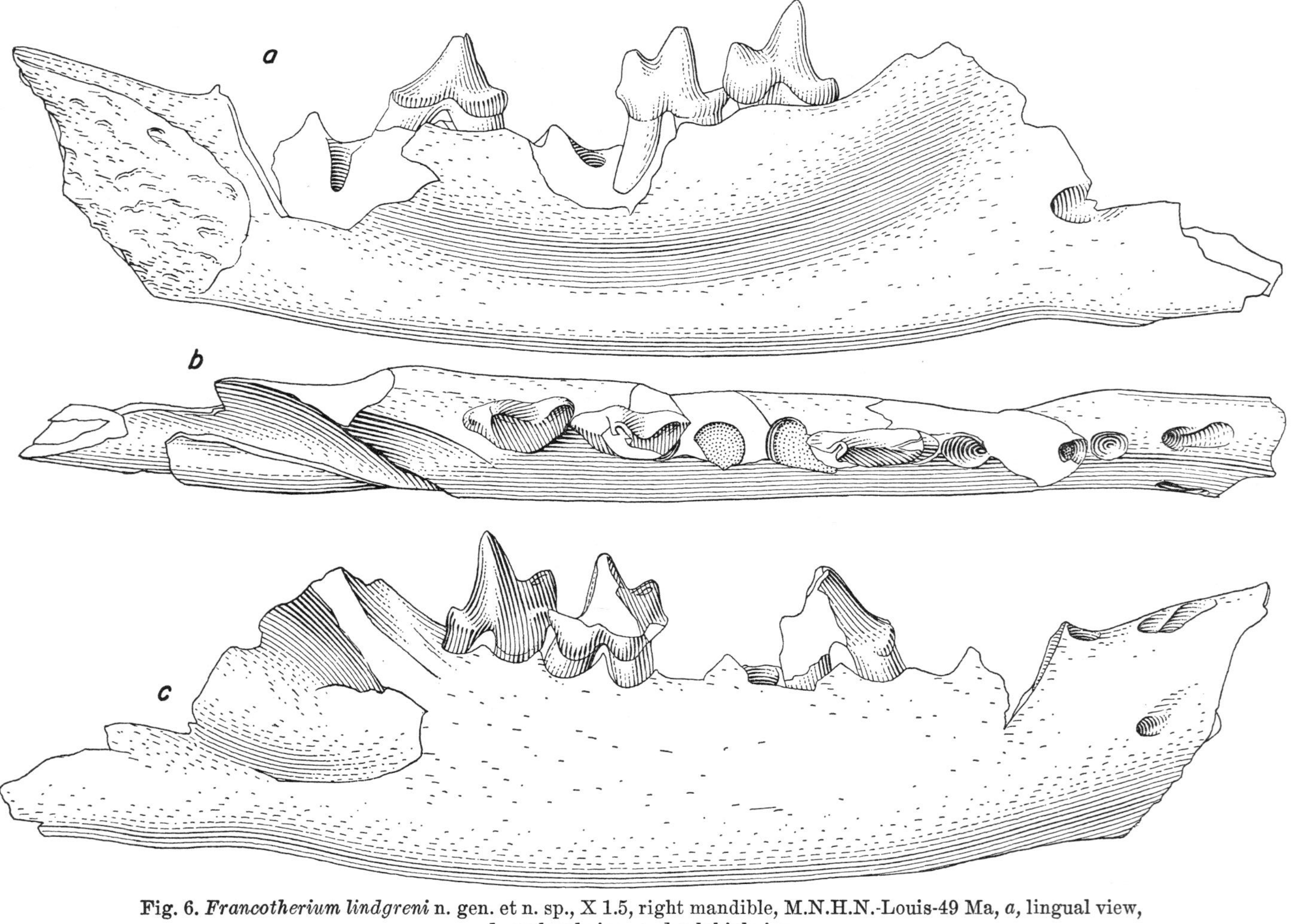

Fig. 6. *Francotherium lindgreni* n. gen. et n. sp., X 1.5, right mandible, M.N.H.N.-Louis-49 Ma, *a*, lingual view, *b*, occlusal view, and *c*, labial view.

sis Matthew and Granger, 1925*d*, (A.M.N.H. 20128, see Matthew and Granger, 1925*d*, fig. 3) is the species with the closest resemblance to this specimen. It is from the late Eocene of Mongolia and differs in that its M_1 is only four-fifths the length of M_3, its P_1 is single-rooted, and its jaw is deeper relative to its length. A fourth difference, contrary to the statement of Matthew and Granger (1925*d*:4), is the presence of a reduced metaconid on the M_3 of *P. irdinensis*. *?Propterodon minutus* from the early Oligocene of Montana differs in that its M_1 is less than three-fifths as long as its M_3 (Douglass, 1901*b*:19).

Van Valen (pers. comm., 1967) has suggested that *Francotherium* and *Prodissopsalis* are the most probable ancestors of the *Hyaenodon-Pterodon* complex.

cf. Subfamilies Limnocyoninae Wortman, 1902*a*, or Hyaenodontinae

TABLE 11
MEASUREMENTS OF LIMNOCYONINAE M^2 OR HYAENODONTINAE M^3

Specimen	Locality	Maximum length	Maximum width
M.N.H.N.-Av 5917 Right ..	Avenay	3.0	7.0

DESCRIPTION
(Fig. 7)

The general outline of this most posterior upper molar in occlusal view is an obtuse triangle with the major angle at the metastylar corner. The anterior border is straight, the posterior border slightly concave, and the labial border convex. The paracone is much higher and larger than the metacone and is located anterolabial to it. The labial height of the paracone is slightly greater than the maximum length of the tooth. The protocone is lower and smaller than the metacone; it is located about one-eighth the width of the tooth lateral to the lingual border and

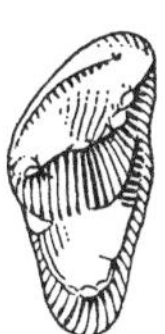

Fig. 7. Limnocyoninae right M^2 or Hyaenodontinae right M^3, X 3, M.N.H.N.-Av 5917, occlusal view.

is lingual to the paracone. The tips of the three major cusps are rounded. The preprotocrista extends only a short way anterolabially from the protocone before it is terminated by a carnassiform notch. The area immediately labial to this notch has been either abraded or damaged. If a paraconule was present, it was not a prominent cusp. The paracingulum extends along the anterior edge of the tooth from the abraded area to a point on the labial side of the paracone base. The postprotocrista extends from the protocone nearly to the metacone and is separated from it by a broad V-shaped notch. The paracrista extends from the paracone to the anterolabial corner of the tooth with a few swellings along its length. There is no parastyle, metacrista, or metastyle. The ectocingulum is continuous along the labial margin of the tooth and has only a few small swellings and no distinct cusps. The paracone is joined to the metacone by a short centrocrista; the bases of the paracone and metacone are connate.

COMMENTS

This tooth could be a hyaenodontine M^3 or a limnocyonine M^2. In comparison with an M^3 of *Prototomus multicuspis* (A.M.N.H. 14773, see Matthew, 1915*d*, fig. 72), the Avenay specimen is similar in the presence of a small metacone; the prominence of the parastylar spur extending labially from the body of the tooth; the arrangement of the paracrista, preprotocrista, and postprotocrista; and the placement of the protocone. In contrast, however, the base of the metacone on the M_3 of *P. multicuspis* bulges posteriorly to a greater degree; the metacone is posterior rather than posterolingual to the paracone; and the outline of the M^3 is narrower anteroposteriorly relative to its width.

This specimen is similar to the M^2 of *Thinocyon medius* from the Bridgerian of Wyoming figured by Denison (1938, fig. 10) in all the characters mentioned in the above paragraph is showing resemblance to the M_3 of *Prototomus multicupis*. In addition to those characters, the outline in occlusal view of the two specimens is quite similar. They differ in that the base of the metacone of *T. medius* does not bulge as far posteriorly, and, as in *P. multicuspis,* its metacone is posterior rather than posterolingual to the paracone.

This tooth may belong to a primitive member of either the Limnocyoninae or Hyaenodontinae.

cf. Superfamily Hyaenodontoidea

TABLE 12

MEASUREMENTS OF HYAENODONTOIDEA M^3

Specimen	Locality	Maximum length	Maximum width
M.N.H.N.-Av 5703 Left....	Avenay	2.4	...

DESCRIPTION
(Fig. 8)

This specimen consists of a labial fragment of the most posterior upper molar. It is complete to an anteroposterior line drawn slightly lingual to the paracone and metacone bases. The fragment indicates that the general outline of the complete tooth in occlusal view was an obtuse triangle with the largest angle at the metastylar corner. The parastylar spur extends well away from the main body of the tooth. The paracone is higher and much larger in basal dimensions than the metacone and is located anterior to it. The tips of both cusps have been flattened by abrasion. A cingulum is developed along the anterior side of the paracone base. The paracrista extends from the paracone to the small parastyle on the anterolabial corner of the tooth. The labial end of the metacingulum is preserved immediately lingual to the metacone from which it is separated by a notch. The ectocingulum is developed all along the labial margin of the tooth except in the metastylar area. The weak parastyle is the only cusp on the ectocingulum. Abrasion on the paracone and metacone has destroyed all sign of a centrocrista but if one were present it must have been short for the bases of the cusps are in juxtaposition.

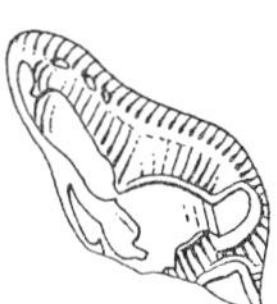

Fig. 8. Hyaenodontoidea left M³, X 6, M.N.H.N.-Av 5703, occlusal view.

COMMENTS

The connate nature of the paracone and metacone on this tooth suggest affinity with the deltatheridians. The following combination of characters also indicates a relationship to the deltatheridians: elongated parastylar spur, wide stylar shelf, and lack of a mesostyle. Of all the deltatheridians, the teeth of the primitive hyaenodontoids show the greatest similarity to this tooth.

TABLE 13

MEASUREMENTS OF OXYAENA MENUI

Specimen				Locality	
M.N.H.N.-1937-586 (Cuis) Left mandible				Cuis	
	Width of posterior root at alveolar border			Length of posterior root at alveolar border	
P₄............	5.2			4.5	
	Width of anterior root above alveolar border		Width of posterior root at point of contact with enamel	Total length at base of root	
M₁..........	6.8		6.4	10.2	
	Trigonid width	Trigonid length	Talonid width	Total length	Length at top of roots
M₂..........	7.7	8.8	4.5	12.1	10.8±.2

Superfamily Oxyaenoidea Osborn, 1910b
Family Oxyaenidae Cope, 1877k
Subfamily Oxyaeninae Trouessart, 1885a
Oxyaena Cope, 1874o

Oxyaena menui, new species

Etymology.—Named in honor of M. Menu who greatly assisted in the field work.

Type.—M.N.H.N.–1937–586 (Cuis), fragment of left mandible with complete M_2, roots of M_1, and posterior root of P_4.

Type locality.—Cuis, Sables-à-Térédines, Cuisian (latest early Eocene), Paris Basin.

Referred specimens.—None.

Diagnosis.—M_2 differs from that of *Oxyaena gulo* Matthew, 1915*d*, in the following ways: trigonid broader relative to total tooth length; talonid shorter and narrower relative to trigonid length and width respectively; hypoconid absent; and crista obliqua weak.

DESCRIPTION

(Fig. 9*a–d*)

The mandible is deep (34.2 mm, measured on the lingual side just anterior to the M_3) and thick (12.3 mm at a point directly below the M_1 halfway to the ventral border). The masseteric fossa is deep: a layer of bone only 2.5 mm thick forms the medial surface of the fossa in contrast to the surrounding bone which is 12 mm thick. The fossa is separated from the ventral border of the mandible by a broad ridge and from the posterior part of the tooth row by a much narrower ridge which is oriented posterodorsally. About 10 mm beneath the tooth row on the lateral side is a broad depression extending anteroposteriorly.

M_2—The trigonid of the M_2 is somewhat longer than wide. The paraconid is slightly higher and much larger in basal dimensions than the metaconid and is located anterior and somewhat lateral to it. The paraconid inclines forward slightly, and its base is anterior to the mesial root of the molar. A small distinct cusp is developed on the anterior side of the paraconid near the base of the crown. The labial height of the protoconid is one and one-third the trigonid length. It is the tallest cusp on the tooth and is subequal to the paraconid in basal dimensions. The tip of the protoconid is damaged but seems to have been as sharp as the metaconid and paraconid. A low paralophid and protolophid each with a carnassiform notch, form the anterior and posterior sides of the prefossid. The prefossid is deep but completely open on the lingual side of the tooth, for the paraconid is markedly anterior to the metaconid with a significant gap between them.

Nearly all trace of the anterior cingulum has been obscured by wear, but enough remains to indicate that it was located on the labial half of the prevallid and dipped laterally downward at an angle of 60 degrees.

The talonid is two-thirds as wide as it is long. Its width is three-fifths and its length three-eighths that of the trigonid. No trace of a distinct hypoconid is present. Separated from the posterior wall of the trigonid by a narrow notch, the entocristid and postcristid form an otherwise continuous, smoothly curving, posterolingually convex ridge from the metaconid to the midpoint of the posterior border of the talonid. Along this ridge are three similar, small, closely spaced cusps that may include the homologues of the hypoconulid and entoconid. At the midpoint of the posterior border of the talonid, the postcristid abruptly decreases in height and extends from there in an anterolabial direction. It passes into the

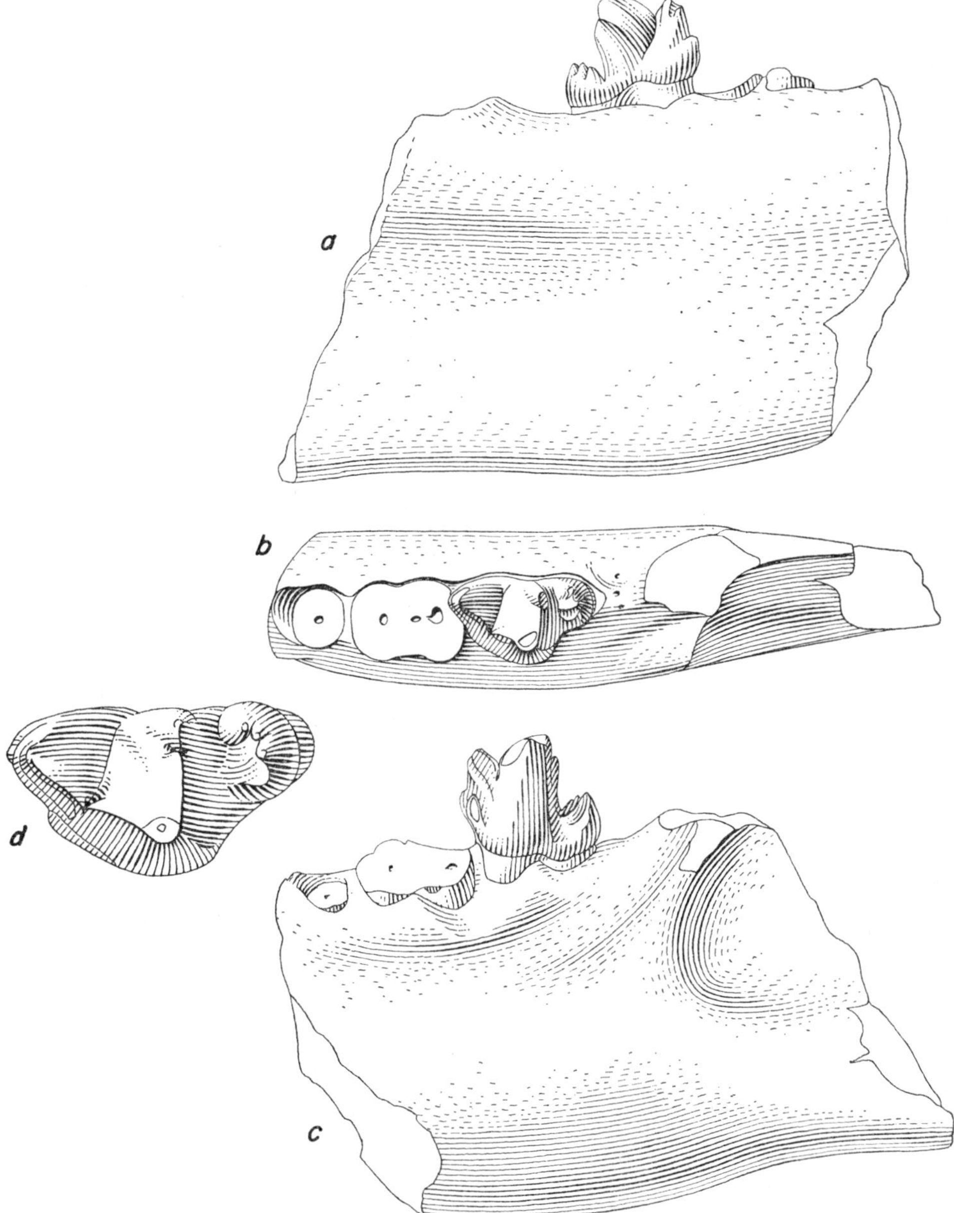

Fig. 9. *Oxyaena menui* n. sp., M.N.H.N.-1937-586 (Cuis). Left mandible, X 1.5, *a*, lingual view, *b*, occlusal view, and *c*, labial view. Left M₂ from same jaw illustrated at larger scale, X 3, *d*, occlusal view.

crista obliqua which is a weak ridge extending anterolingually toward the metaconid. A small shallow postfossid is developed on the lingual half of the talonid. There is no indication of an ectostylid. Prominent, nearly flat shearing surfaces are present on the anterior and posterior faces of the trigonid, covering the surface lateral to an imaginary line between the tips of the lingual trigonid cusps and the exterior of the tooth at the base of the enamel.

P_4 and M_1—Some inferences about the size of the P_4 and M_1 can be made from what remains of the roots of these teeth. The trigonid was only slightly wider than the talonid on M_1. Compared with the M_2, the M_1 was shorter, had a wider talonid, and an equally wide trigonid. The width of the talonid on P_4 was greater than that on M_2 and either equal to or less than that on M_1.

COMMENTS

Among the trends of evolution of M_2 within *Oxyaena* are lengthening of the prevallid, increase of the angle between the prevallid and postvallid, and reduction of the metaconid and talonid (cf. Denison, 1938:197). The morphology of the M_2 of *O. menui* indicates that it is at a stage of evolution closest to that of the Lower Graybull species *O. gulo* but shows some similarities with the Upper Graybull to Lost Cabin, *O. forcipata*.

Oxyaena menui and *O. gulo* (A.M.N.H. 15193, see Matthew, 1915*d*, fig. 48) have about the same angle between the prevallid and postvallid, and an equally prominent metaconid on the M_2. The M_2 of *O. menui* differs from that of *O. gulo* in the markedly shorter and slightly narrower talonid relative to the trigonid length and width respectively, absence of a hypoconid, weaker crista obliqua, and greater relative breadth of the tooth (trigonid width: total length). *O. menui* is closer in size to *O. gulo* than to *Oxyaena forcipata* (A.M.N.H. 15183, see Matthew, 1915*d*, fig. 46). The M_2 of *O. forcipata* resembles that of *O. menui* in the reduced length of the talonid. The M_2 of *O. forcipata* is somewhat broader than that of *O. gulo* and hence closer to the breadth of *O. menui*. The M_2 of *O. menui* differs from that of *O. forcipata* in the following characters: smaller angle between the prevallid and postvallid, greater strength of the metaconid, narrower talonid, absence of a hypoconid, and weaker crista obliqua.

The ratio of the lengths of the M_1 and M_2 of *Oxyaena menui* is not an acceptable comparative datum because of the great magnitude of the probable error in the estimate that was made. Comparison of the mandibles merely reaffirms that *O. menui* is closer in size to *O. gulo* than to *O. forcipata*.

In summary, the characters of the M_2 indicate that *Oxyaena menui* is closer to *O. gulo* in overall morphology and size and to *O. forcipata* in the reduced length of the talonid and greater width of the trigonid. A weak crista obliqua is known to occur in North American species of *Oxyaena*. *O. ultima* (see Denison, 1938, fig. 1) from the Wasatchian of Wyoming, a species different from *O. menui* in many other respects, also has a weak crista obliqua.

Denison (1938:247–248) proposed a phylogenetic dichotomy in *Oxyaena* based on tooth size and width. *O. gulo* was part of a lineage characterized by small narrow teeth and *O. forcipata* was in a line with large broad teeth. Van Valen, who recently reviewed the group, concurred in this view (1966:79). It is doubtful whether *O. menui* can be fitted into such a lineage with the evidence now available. Though the tooth is wide, it is small and thus does not fit neatly into Denison's dichotomy. It seems possible, therefore, that *O. menui* represents a lineage that differentiated from the North American stock at about the time when the *O. gulo* and *O. forcipata* lineages were differentiating (Tiffanian = late Paleocene, North America). If this is the case, it is more likely that the *O. gulo* lineage was ancestral because the earliest *O. gulo* is smaller than *O. menui*; this is not true of the *O. forcipata* lineage. Data presented by Denison (1938) indicate that there is no instance where the dentition became smaller in a North American lineage of *Oxyaena*.

An M_2 tentatively referred to *Oxyaena* has been previously reported from England. This tooth was originally described and figured by Cooper (1932*b*:464; fig. 2; pl. 12, fig. 4) and compared with *Sinopa*. Van Valen refigured and redescribed the tooth, tentatively referring it to *Oxyaena* (1965*b*:658–659, fig. 5*a*). The M_2 of *O. menui* differs in the narrowness of the talonid relative to the trigonid width, and in the absence of a hypoconid.

Oxyaena sp.

TABLE 14
MEASUREMENTS OF OXYAENA SP. M_1

Specimen	Locality	Trigonid width	Trigonid length	Talonid width	Total length
U.C.M.P. 83754 Left.....................	Sinceny	7.1	7.5	7.0	11.6

DESCRIPTION
(Fig. 10*a–c*)

This M_1 from Sinceny has a trigonid slightly longer than wide. The paraconid is somewhat lower and equal in basal dimensions to the metaconid and is located anterior and slightly lateral to it. The paraconid inclines forward slightly, and its base is forward of the anterior root of the tooth. A sharp distinct ridge begins at the apex of the paraconid and extends in a posteroventral direction down the posterolingual side of the cusp for about two-fifths the distance to the base of the crown. The protoconid is the highest cusp on the tooth, its labial height above the talonid is slightly less than the length of the trigonid. The apices of all trigonid cusps are quite sharp, but the paraconid has a slight indication of horizontal truncation of the cusp. A low paralophid and protolophid, each with a carnassiform notch, form the anterior and posterior sides of the prefossid. The prefossid is deep and completely open on the lingual side of the tooth, for the paraconid is well separated from the metaconid.

A wide anterior cingulum dipping downward laterally at about 30 degrees is present on the labial two-thirds of the prevallid. An appression fossette is medial and adjacent to the lingual end of the anterior cingulum.

The talonid is three-fifths as long as it is wide. It is equally as wide as the trigonid but only slightly more than half as long. The well-developed hypoconid is the only talonid cusp. It is located at a point slightly labial to the center of the talonid and is markedly lower than the entocristid. The entocristid and postcristid form a high, continuous, smoothly curving, posterolingually convex ridge from the metaconid to the midpoint of the posterior border of the talonid. There the postcristid starts to descend to the hypoconid. Meeting the postcristid at the point

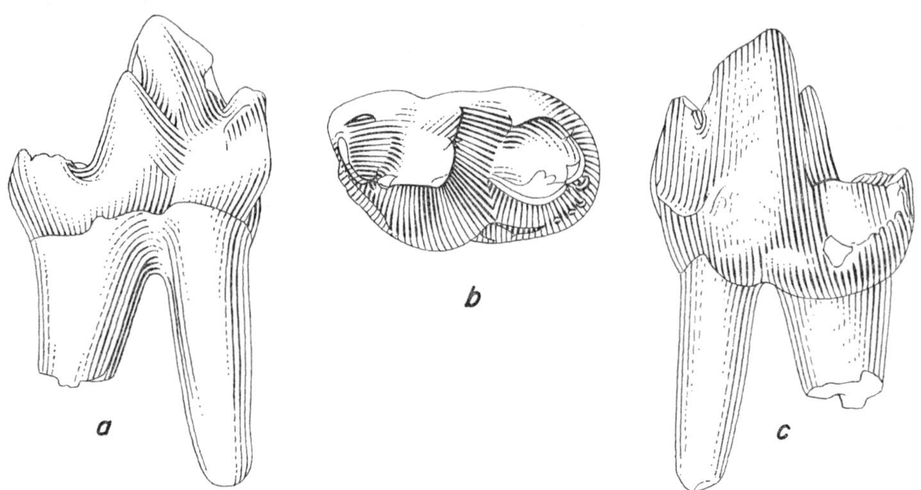

Fig. 10. *Oxyaena* sp., X 3, left M₁, U.C.M.P. 83754, *a*, lingual view, *b*, occlusal view, and *c*, labial view.

where it begins to descend toward the hypoconid, a labial cingulum extends along the lateral side of the talonid to the posterolateral corner of the trigonid.

COMMENTS

This M_1 and the M_2 of *Oxyaena menui* described above have many of the relationships present between the M_1 and M_2 of *O. forcipata* (A.M.N.H. 4209, see Cope, 1884*o*, pl. 24*c*, fig. 11). In both instances, the M_1 in comparison to the M_2 is a shorter tooth with a smaller trigonid and narrower angle between the prevallid and postvallid. However, unlike *O. forcipata*, the M_1 talonid of cf. *Oxyaena* is greater in length and width relative to that on the M_2 of *O. menui*. The roots of M_1 preserved on the type specimen of *O. menui* indicate that the complete tooth was approximately the same length and width as this isolated M_1.

This M_1 shows a similar quantitative relation to the M_2 of *Oxyaena* sp. from England mentioned above and described by Cooper (1932*b*:464; fig. 2; pl. 12, fig. 4) and Van Valen (1965*b*:658–659, fig. 5*a*) as it does to the M_2 of *O. menui*. The talonids of this M_1 and the M_2 of *Oxyaena* sp. show an additional similarity in that the hypoconid is low and the crista obliqua strong. In contrast, the M_2 of *O. menui* has a weak crista obliqua and lacks a hypoconid.

cf. Superfamily Oxyaenoidea or Hyaenodontoidea

DESCRIPTION
(Fig. 11a–c)

The specimen consists of a left mandible fragment with a single tooth (P_4) preserved. It retains the alveoli for C-M_1. The depth of the mandible immediately anterior to the P_4 is 25 mm; its width at that point is 12 mm. There are two mental foramina present; the larger is about 10 mm below the point of contact between the P_2 and P_3 and elongated anteroposteriorly; the smaller one is not quite so far below the anterior edge of the P_1. There is no trace of a dental foramen on the

TABLE 15

MEASUREMENTS OF A JAW FRAGMENT OF OXYAENOIDEA OR HYAENODONTOIDEA

Specimen				Locality	
M.N.H.N.-Louis-68 Gr Left mandible				Grauves	
	Width of anterior alveolus	Length of anterior alveolus	Width of posterior alveolus	Total alveolar length	
Left P_1..............	2.0	2.0	3.2	6.3	
Left P_2..............	4.5	4.1	4.9	7.0	
Left M_1..............	6.9	5.1	...	...	
	Width of anterior root	Length of anterior root	Width of posterior root	Total root length	
Left P_3..............	5.5	4.7	5.5	10.4	
	Trigonid width	Trigonid length	Talonid width	Total length	
Left P_4..............	6.5	6.9	6.8	10.2±.3	

lingual side of the mandible. The nonfused midline symphysis covers all the lingual surface of the mandible anterior to P_2 except for a narrow strip immediately below the P_1.

The elliptical cross section of the canine exposed on the broken anterior edge of the mandible is flattened on the lingual side; its mediolateral diameter is 7.6 mm; its dorsoventral diameter is 11.5 mm.

P_4—The trigonid of the P_4 is subequal in length and width. Because of extensive wear, the original heights of the trigonid cusps cannot be ascertained, but enough of the tooth remains so that conjectures can be made as to the unworn state. The paraconid was lower than the protoconid, somewhat smaller in basal dimensions and located in a position anterior and slightly lingual to it. The base of the cusp is anterior to the mesial root of the tooth and the cusp probably inclined forward somewhat. The protoconid was the highest and largest cusp on the tooth and occupied the entire posterior half of the trigonid except for the low and small metaconid located on the posterolingual corner. No indication of a prefossid was observed. An appression fossette is deeply incised into the anterior edge of the trigonid.

The length of the talonid is less than half its width; it is subequal in width to the trigonid and only half as long. There are two distinct talonid structures: a weak crista obliqua extending from the posterior border of the tooth to the trigonid and the base of a rather large cusp which may be the entoconid. The postfossid is quite shallow and opens lingually. The tooth has surfaces of wear on the dorsal and posterior surfaces of the trigonid and a prominent wear facet on the labial side of the talonid.

COMMENTS

The P_4 morphology suggests that this specimen could be referred to either the Miacidae or the Palaeoryctoidea, but the comparatively great width of the P_4

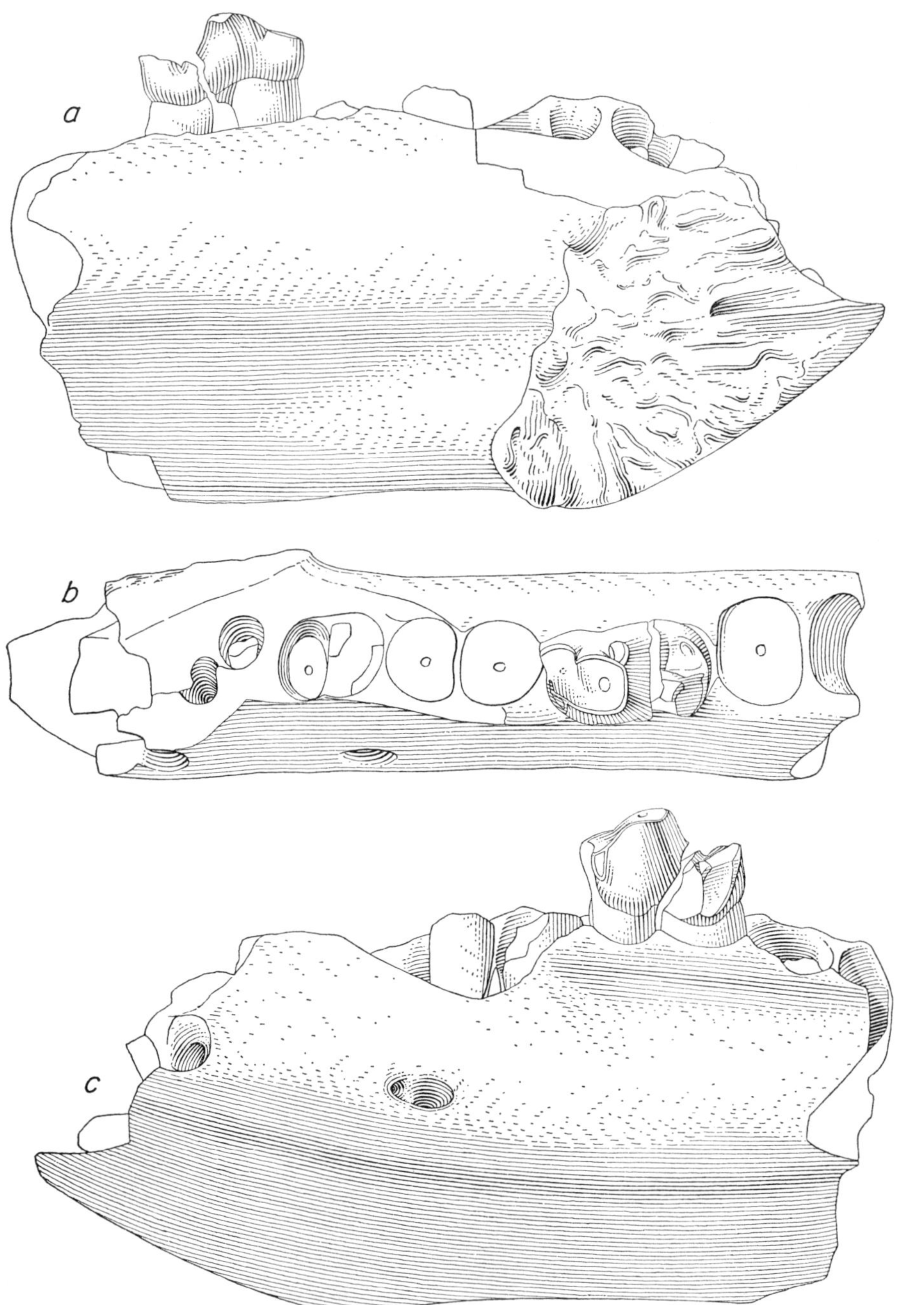

Fig. 11. cf. Oxyaenoidea or Hyaenodontoidea, X 2, left mandible, M.N.H.N.-Louis-68 Gr, *a*, lingual view, *b*, occlusal view, and *c*, labial view.

relative to the M_1 appears to exclude this specimen from the miacids and the relatively large size of the tooth and mandible excludes it from the palaeoryctoids. Thus by elimination, it is referred to the oxyaenoids or hyaenodontoids.

Order CARNIVORA Bowdich, 1821
Suborder FISSIPEDA Blumenbach, 1791
Superfamily Miacoidea Simpson, 1931*b*
Family Miacidae Cope, 1880*c*
Subfamily Viverravinae Matthew, 1909*d*
cf. *Viverravus* Marsh, 1872*g*

TABLE 16

MEASUREMENTS OF CF. VIVERRAVUS P^4

Specimen	Locality	Maximum length	Maximum width
M.N.H.N.-Mu 6216 Left....	Mutigny	4.3	3.0

DESCRIPTION

(Fig. 12*a*)

The enamel of this specimen appears to have been chemically etched, possibly by the digestive fluids of some unknown carnivore or by acids produced through plant decay. The general outline of this tooth in occlusal view is a right triangle with the hypotenuse formed by the nearly straight lingual edge of the tooth. The anterior side is deeply concave and the labial side is slightly concave posterior to the paracone. The parastyle is much smaller in basal dimensions and lower than the paracone and located anterior and slightly lingual to it. The tips of both cusps have been flattened by abrasion. The protocone is subequal to the parastyle in height and basal dimensions, thus lower and smaller than the paracone. It is located on the lingual edge of the tooth equidistant from the paracone and parastyle. As on those two cusps, its tip has been flattened by abrasion. The only crest developed on the tooth is the metacrista, which extends from the paracone to the posterolabial corner of the tooth forming a prominent blade. There is no development of a stylar shelf. The bases of the paracone and parastyle are connate but there is no paracrista.

COMMENTS

The prominent parastyle on this P^4 definitely excludes it from *Miacis*. The specimen is provisionally referred to *Viverravus* because of its high degree of morphological similarity to the much larger P^4 of *V. sicarius* Matthew, 1909*d* (type specimen A.M.N.H. 11521, see Matthew, 1909*d*, figs. 7, 13, and 14) from the Bridgerian of Wyoming. There is no morphological character that excludes the possibility that the tooth properly belongs in the North American genera *Protictis* (Paleocene) or *Ŏodectes* (Eocene).

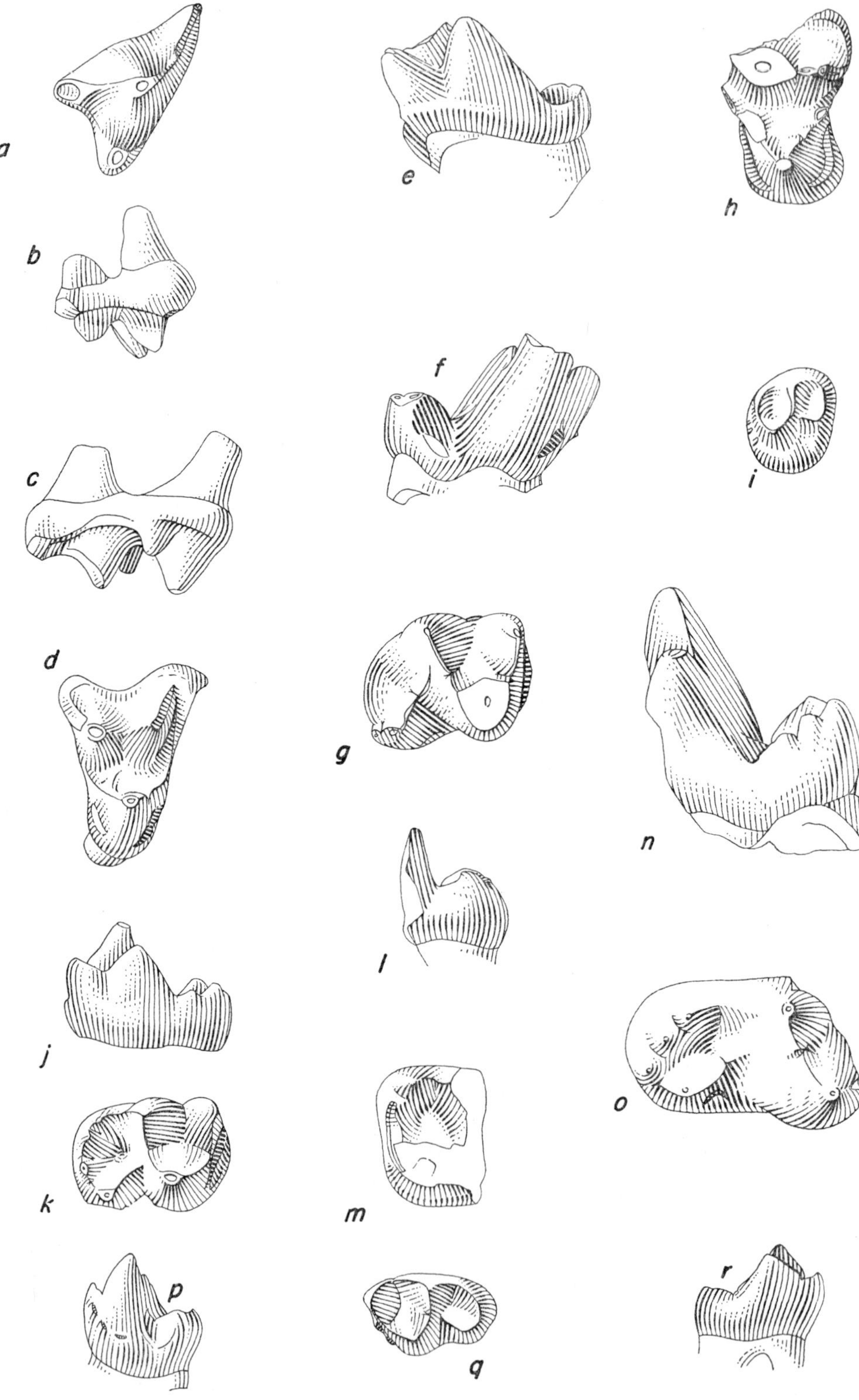

TABLE 17

MEASUREMENTS OF CF. VIVERRAVUS M[1]

Specimen	Locality	Maximum length	Maximum width
M.N.H.N.-Av 4916 Right...	Avenay	3.7	5.0

DESCRIPTION

(Fig. 12b–d)

The general outline of this tooth is an acute angle isosceles triangle with the base formed by the labial edge of the tooth. Impressed on the labial side of the tooth is a deep ectoflexus; the posterior side of the tooth is slightly concave; and the anterior side is straight. Height and basal dimensions of the paracone are greater than those of the metacone. The position of the paracone is anterior and slightly labial to the metacone. The bases of the two cusps are distinct from one another. The protocone is taller and greater in basal dimensions than the metacone and subequal to the paracone in height and basal dimensions. Lingual to a point one-third the distance from the paracone to the metacone, the protocone is one-third the width of the tooth from its lingual edge. A small area at the tip of each cusp is rounded. A narrow precingulum and postcingulum are developed on the lingual third and sixth of the anterior and posterior sides of the tooth respectively. Extending anterolabially from the protocone, the well-developed preprotocrista gives way to the precingulum at the paraconule, a raised section of this otherwise continuous crest. The weaker postprotocrista extends postero-labially and terminates lingual to the metacone, no metaconule or metacingulum being developed. Joining the metacone and paracone is a weak centrocrista. A prominent stylar shelf is present for the entire length of the tooth. Between the paracone and the labial edge of the tooth is a weak paracrista. There is no indication of a parastyle in the anterolabial corner of the tooth. Extending postero-labially from the metacone is a well-worn metacrista. Most of the wear on this tooth is in the labial region. In order of decreasing magnitude, wear is apparent on the metacrista, paracrista, metacone, centrocrista, and protocone.

COMMENTS

The M[1] on A.M.N.H. 15181 that was identified by Matthew (1915*d*, fig. 22) as *Viverravus acutus* is a slightly larger tooth. In outline, the two are similar except that the posterolabial corner is more broadly rounded and the precingulum and postcingulum are wider on *V. acutus*. They agree in every other feature com-

---

Fig. 12, X 6. cf. *Viverravus*, left P[4], M.N.H.N.-Mu 6216, *a*, occlusal view. cf. *Viverravus*, right M[1], M.N.H.N.-Av 4916, *b*, labial view, *c*, anterior view, and *d*, occlusal view. cf. *Uintacyon*, right M[1], M.N.H.N.-Av 5006, *e*, lingual view, *f*, labial view, and *g*, occlusal view. cf. *Miacis*, left M[1]. M.N.H.N.-Mu 6568, *h*, occlusal view. cf. *Miacis*, left M[3], M.N.H.N.-Av 5817, *i*, occlusal view. cf., Miacidae, right lower molar, M.N.H.N.-Av 5708, *j*, lingual view, and *k*, occlusal view. cf. Miacidae, right molar talonid, M.N.H.N.-Mu 6613, *l*, lingual view, and *m*, occlusal view. cf. Miacidae, right lower molar, M.N.H.N.-Av 5012, *n*, lingual view, and *o*, occlusal view. cf. Miacidae, left M[2], M.N.H.N.-Av 5895, *p*, labial view, *q*, occlusal view, and *r*, lingual view.

pared: relative position, basal dimensions, and heights of the protocone, meta-cone, and paracone; relative strengths and placement of the crests and paraconule; and absence of a metaconule and parastyle. Because the differences mentioned above are so slight and other species of *Viverravus* show similarity to the French specimen in these features (cf. M^1 of *V. sicarius,* A.M.N.H. 56511, see Robinson, 1966, pl. 5, fig. 12), there is no serious doubt as to the proper generic assignment for this specimen.

Subfamily Miacinae Trouessart, 1885*a*
cf. *Miacis* Cope, 1872*e*

TABLE 18

MEASUREMENTS OF CF. MIACIS M^1

Specimens	Locality	Maximum length (teeth damaged)	Maximum width (teeth damaged)
M.N.H.N.-Mu 6568 Left....	Mutigny	3.2	4.9
M.N.H.N.-Av 4756 Right...	Avenay	2.5	4.2

DESCRIPTION

(Fig. 12*h*)

The general outline of these teeth in occlusal view is a trapezoid with the base formed by the labial side. The labial side is damaged except posterior to the meta-cone on the best preserved specimen; the lingual side is convex; the postvallum is concave and the prevallum is straight where preserved, having been damaged labial to the paracone on both specimens. The tip of the paracone is missing but enough remains to show that it is a larger and higher cusp than the metacone. The paracone is anterior and slightly labial to the metacone. The protocone is lower and larger in basal dimensions than the metacone and subequal to the paracone in basal dimensions. Located about one-fourth the width of the tooth lateral to its medial edge, the protocone is lingual to a point about one-fourth the distance from the paracone to the metacone. The apex of the protocone is rounded, and that of the metacone is sharp. A wide precingulum is developed on the lingual half of the prevallum and an equally wide postcingulum on the lingual third of the postvallum. A well-developed preprotocrista extends anterolabially from the protocone to the paraconule at the anterior border of the tooth. A well-developed paracingulum extends along the anterior side of the base of the paracone from the paraconule to the labial damaged area. The postprotocrista is well developed and extends to the metaconule which is a mere swelling on that crest. From there the equally prominent metacingulum extends to a point posterior to the metacone. The metacrista extends posteriorly from the metacone to the posterior side of the tooth and there turns labially towards the metastylar region. A short centrocrista joins the paracone and metacone. There is a broad stylar shelf labial to the meta-cone with a small distinct ectocingulum extending forward from the metastylar region for a short way. It is cut off by the extensive damage at the anterior part of the stylar shelf.

COMMENTS

The anteroposteriorly expanded lingual region of this tooth immediately separates it from the deltatheridians. It resembles most closely the M^1 of *Miacis* among the upper cheek teeth of miacids. Direct comparison with specimens of *Miacis exiguus* (A.M.N.H. 15176, see Matthew, 1915*d*, fig. 28) from the Wasatchian of Wyoming and *Miacis sylvestris* (A.M.N.H. 13071, see Matthew, 1909*d*, fig. 15) from the Bridgerian of Wyoming reveals that the North American forms are one-third larger. All these *Miacis* specimens share the following common characters: well-developed precingulum and postcingulum; connate paracone and metacone; morphology and relative sizes of the protocone, paracone, metacone, and paraconule as described above for the French specimen; outline of posterior border of tooth in occlusal view concave; wide stylar shelf opposite the metacone with a slight ectocingulum at the labial edge of the tooth; and absence of a hypocone. These features taken together distinguish the M^1 of these *Miacis* specimens from the other genera of miacids.

TABLE 19

MEASUREMENTS OF CF. MIACIS M₃

Specimens	Locality	Greatest width	Greatest length
M.N.H.N.-Av 5817 Left....	Avenay	2.5	2.3
M.N.H.N.-Av 4969 Right...	Avenay	2.2	1.9
M.N.H.N.-Av 5820 Left....	Avenay	1.9	2.1
M.N.H.N.-Av 4934 Right...	Avenay	2.2	2.2
M.N.H.N.-Av 229 Left.....	Avenay	2.3	2.5

DESCRIPTION

(Fig. 12i)

These extremely low and small button-shaped teeth possess a trigonid with the cusps arranged in an equilateral triangle and ranging in degree of development from mere swellings on the paralophid and protolophid to quite prominent structures. The cusps are subequal in height. In basal dimensions, the metaconid is the largest, the paraconid intermediate, and the protoconid the smallest although the degree of difference is not great. There is a prominent expansion on the anteroexternal side of the tooth which on some specimens bears accessory cusps or an anterior cingulum. The prefossid is shallow and bordered anteriorly and posteriorly by the paralophid and protolophid, but the lingual side is open between the widely separated paraconid and metaconid.

The talonid is variable with the crista obliqua ranging from completely absent to a prominent structure extending from the faint hypoconid nearly to the protolophid. The postcristid is always present and may even continue labially beyond the point of contact with the crista obliqua.

COMMENTS

The morphological heterogeneity of this group of teeth may be due to one or both of the following: (1) the teeth belong to different species or genera, and (or) (2) the M_3 is of lesser functional importance in miacids than the M_1 and M_2 and therefore a greater variation may be expected.

The near equality of the trigonid and talonid in both height and length suggests affinities with *Miacis*.

cf. *Uintacyon* Leidy, 1872*j*

TABLE 20

MEASUREMENTS OF CF. UINTACYON M_1

Specimens	Locality	Trigonid width	Trigonid length	Talonid width	Total length
M.N.H.N.-Av 5005 Right................	Avenay	3.3	3.2	2.3	4.4
M.N.H.N.-Av 5006 Right...............	Avenay	3.3	3.0	2.7	4.5

DESCRIPTION

(Fig. 12*e–g*)

The length of the trigonid is less than the width. The paraconid is lower than the metaconid. The two cusps are subequal in basal dimensions and well separated from one another. The paraconid is located anterior and somewhat lateral to the metaconid.

The protoconid tip is missing in both specimens. This cusp is larger than either of the other trigonid cusps and is equidistant from them. The apices of the preserved trigonid cusps are dull points. The prefossid is moderately deep with a well-developed paralophid and protolophid (both of which have a prominent carnassiform notch) forming the anterior and posterior boundaries respectively. The prefossid is open on the lingual side of the tooth between the widely spaced metaconid and paraconid. The anterior cingulum is developed on the middle third of the prevallid and dips laterally 30 degrees below the horizontal on M.N.H.N.-Av 5005 and 50 degrees on M.N.H.N.-Av 5006.

The talonid is only half as long as it is wide. It is one-third to one-half as long as the trigonid and three-fourths to nearly as wide. The entocristid and postcristid form an unbroken, posterolingually convex ridge from the hypoconid to a point posterior to the base of the metaconid. The ridge totally lacks any evidence of intermediate cusps. The crista obliqua is convex laterally and meets the trigonid labial to a point just below the carnassiform notch about midway between the protoconid and metaconid. The doubled hypoconid is merely a pair of small swellings on the talonid cristid and is located at the posteroexternal corner of the tooth. The posterior member of this doubled cusp may be the hypoconulid in which case it is somewhat smaller than the hypoconid. The postfossid is shallow and opens lingually between the base of the metaconid and the anterior edge of the entocristid. Between the hypoconid and protoconid there is no evidence of an ectostylid. Shearing surfaces are present on the twinned hypoconid and on a tiny area of the posterior side of the metaconid.

COMMENTS

The trenchant talonid is a character of some miacids and distinguishes these teeth from the hyaenodontids and oxyaenids. The distinct separation between the paraconid and metaconid bases and the low angle between the prevallid and postvallid is the basis for the referral of these teeth to *Uintacyon,* a genus previously known only from the Eocene of North America, rather than to the other miacids that have a trenchant talonid.

These specimens resemble the M_1 of *Uintacyon massetericus* (Cope, 1882e) (type specimen A.M.N.H. 4250, see Cope, 1884o, pl. 24e, fig. 11) from the Wasatchian of Wyoming, in relative heights and arrangement of the trigonid cusps, development of the paralophid and protolophid, the wide lingual prefossid opening between the metaconid and paraconid, development of the anterior cingulum, the angle formed between the posterolingual edge of the trigonid and the dorsal edge of the entocristid, and size of the talonid relative to the trigonid. These specimens differ in that they are only two-thirds the size of *U. massetericus* and the crista obliqua extends anterointernally from the hypoconid rather than anteriorly.

cf. Miacidae

TABLE 21

MEASUREMENTS OF CF. MIACIDAE M_2

Specimen	Locality	Trigonid width	Trigonid length	Talonid width	Total length
M.N.H.N.-Av 5895 Left..................	Avenay	1.8	1.6	1.5	2.9

DESCRIPTION

(Fig. 12p–r)

The length and width of the trigonid of this lower molar are nearly the same. The paraconid is markedly lower and smaller than the protoconid or metaconid and is located anterior and somewhat labial to the latter cusp. The protoconid is slightly higher but equal in basal dimensions to the metaconid, and its labial height above the talonid is about three-fourths the trigonid length. Whereas the apices of the protoconid and metaconid are rounded, that of the paraconid is a blunt point. The protoconid is labial to a point nearly halfway from the metaconid to the paraconid and is one-fourth the trigonid width from the labial edge of the tooth. The apices of the trigonid cusps form an equilateral triangle; the well-developed protolophid, paralophid, and metacristid form the somewhat outwardly convex sides around a deep prefossid. The anterior cingulum on the prevallid dips downward laterally at an angle of 45 degrees and extends from a point one-third the distance from the paraconid to the protoconid to another point at the anteroexternal corner of the tooth.

The length of the talonid is about three-fourths its width. Its width is five-sixths and its length three-fourths that of the trigonid. The entocristid and postcristid form an unbroken arc along the talonid margin from the base of the metaconid to the hypoconid. The somewhat lingually concave crista obliqua extends from the hypoconid to a point at the base of the trigonid equidistant from the paraconid and metaconid. The hypoconid, hypoconulid, and entoconid are mere swellings on the cristids located respectively (1) one-fourth the talonid width from the labial border, (2) midway between the labial and lingual talonid borders on the posterior edge of the tooth, and (3) at the posterointernal corner of the tooth medial to the first-mentioned cusp. The postfossid is well enclosed and deep. A faint ectostylid is present between the protoconid and hypoconid. No obvious wear facets are present on this tooth.

COMMENTS

The presence of a well-developed metacristid excludes this tooth from inclusion in the Deltatheridia.

Many features in common between the M_2 on A.M.N.H. 11495, type specimen of *Őodectes proximus* (see Matthew, 1909*d*, figs. 7 and 8), and this specimen suggest that it might be referable to that genus. Features that they share include the position, relative heights, and basal dimensions of the trigonid cusps; strength and position of the anterior cingulum; strengths of the protolophid, paralophid, and metacristid; height of the trigonid; relative lengths and widths of the trigonid and talonid; position of the entocristid, postcristid, and crista obliqua on the talonid; degree of enclosure of the prefossid and postfossid; and weakness of the talonid cusps. Though not as numerous, the differences between the two indicate that the French specimen may belong in another genus, perhaps non-miacid. Immediately behind the trigonid, the entocristid on the M_2 of *Ő. proximus* is low and becomes higher posteriorly. On the French specimen, the entocristid is high for its full extent. A second difference is that the protolophid, paralophid, and metacristid are straight on the M_2 of *Ő. proximus* rather than bowed outward from the center of the trigonid. Although the former feature can be found in other miacid genera (cf. M_2 of *Vulpavus australis*), the latter to my knowledge does not occur in any known miacid. For this reason, the familial assignment of this specimen is provisional.

TABLE 22

MEASUREMENTS OF CF. MIACIDAE LOWER MOLAR

Specimen	Locality	Trigonid width	Trigonid length	Talonid width	Total length
M.N.H.N.-Mu 6613 Right..............	Mutigny	...	...	3.6	...

DESCRIPTION

(Fig. 12*l, m*)

The specimen consists of the posterior half of the metaconid and a nearly complete talonid slightly damaged on the occlusal surface in the region of the hypoconid.

The height of the metaconid above the contact between the crown enamel and the root is about three-fourths the width of the talonid. The entocristid extends posteriorly from the base of the metaconid to the entoconid at the posterolingual corner of the tooth. The postcristid extends from the entoconid along the posterior margin of the tooth to the hypoconid. Exactly where the hypoconid and the crista obliqua were cannot be determined owing to the damage alluded to previously, but the crista obliqua definitely abutted against the talonid close to the base of the protoconid. Of the two talonid cusps, the hypoconid was the taller and has the larger basal dimensions. Though smaller, the entoconid is prominent. The post-fossid was shallow but completely enclosed. A labial cingulum is present immediately anterolabial to the hypoconid region. Its development anterior to this region cannot be determined because of the damage sustained by this specimen. There is a slight development of a cingulum posterior to the hypoconid.

COMMENTS

This specimen has many features which are commonly found on the M_1 and M_2 of those primates where the trigonid is distinctly higher than the talonid [cf. the M_1 and M_2 of *Pelycodus ralstoni* (A.M.N.H. 16096, see Matthew, 1915*f*, fig. 6)]. These similarities include the outline of the talonid in occlusal view with its sharp corners and high width to length ratio, the well-developed entoconid, the weak or absent hypoconulid, the strong hypoconid, the labial position of the point of intersection between the crista obliqua and posterior wall of the trigonid, and the presence of labial cingula. However, the vertical orientation of the posterolingual corner of the trigonid of this specimen contrasts greatly with the anteriorly inclined corner on the M_1 and M_2 of these primates.

This specimen resembles the M_1 of *Protictis* (*Bryanictis*) *microlestes* (Simpson, 1935*g*) from the medial Paleocene of Montana in every way it resembles the M_1 and M_2 of the primates mentioned above (cf. MacIntyre, 1966, pl. 17 figs. 1, 3). However the similarity is even closer, for the posterolingual corner of the trigonid of this miacid is also vertical (cf. MacIntyre, 1966, pl. 13, fig. 1).

TABLE 23

MEASUREMENTS OF CF. MIACIDAE LOWER MOLAR

Specimen	Locality	Trigonid width	Trigonid length	Talonid width	Approximate talonid length
M.N.H.N.-Av 5012 Right.............	Avenay	3.6	...	3.6	2.8

DESCRIPTION

(Fig. 12n, o)

Owing to the absence of that part of the tooth anterior to the protoconid and metaconid, nothing can be stated about the trigonid length or the nature of the paraconid. The protoconid is markedly higher and greater in basal dimensions than the metaconid, and though it cannot be compared with the trigonid length, it is quite high relative to the available dimensions of the tooth. The protoconid is located one-fourth the trigonid width lingual to the lateral border of the tooth and is slightly anterior and lateral to the metaconid. The apices of the preserved trigonid cusps are rounded. The prefossid is deep, opens laterally between the metaconid and paraconid, and is bounded posteriorly by a protolophid with a prominent carnassiform notch.

The talonid is three-fourths as long as it is wide and is subequal to the trigonid in width. The postcristid and entocristid extend anterointernally from the hypoconulid straight to a point on the lingual edge of the tooth slightly posterior of the metaconid base. The other arm of the postcristid (labially convex) links the hypoconid and hypoconulid. The crista obliqua (also labially convex) extends to a point at the base of the trigonid equidistant from the protoconid and metaconid and directly below the carnassiform notch. Of the four talonid cusps, the hypoconid is the highest and has the greatest basal dimensions. It is located one-fourth the talonid width lingual to the labial side of the tooth. The hypoconulid, entoconid, and entoconulid are equal in height and basal dimensions and are equally spaced on a straight line with the first and last named cusps at the middle of the posterior and lingual edges of the talonid respectively. The postfossid is moderately deep and open both labially and lingually near the trigonid. No trace of an ectostylid or wear facets is present.

COMMENTS

Superficially this molar is similar to an M_2 of *Prototomus* or *Proviverra* (= *Sinopa*) sp. (U.C.M.P. 43957, see McKenna, 1960*h*, fig. 48) from the Wasatchian of Colorado. The two teeth are similar in basal dimensions, heights of the protoconid and the metaconid, rounding of the metaconid apex, development of a protolophid carnassiform notch, and degree of development and arrangement of the talonid cristids. The above described molar differs from this M_2 of *Prototomus* or *Proviverra* sp. in that its protolophid dips steeply downward from the metaconid to the carnassiform notch rather than being horizontal; its talonid is equal to its trigonid in width rather than being narrower; and it has a well-developed hypoconulid, entoconid, and entoconulid. The last two differences, particularly the latter one, separates this specimen from the lower molars of the other deltatheridians as well as *Prototomus* and *Proviverra* (Van Valen 1966*b*:95).

This molar is similar to the M_1 of *Protictis haydenianus* (A.M.N.H. 3368, 3371, 35390, see MacIntyre, 1966, pls. 6, 7, and 11) discussed by MacIntyre (1966:154–155) in the heights of the protoconid and the metaconid, the rounding of the metaconid apex, the development of a protolophid carnassial notch, the dipping of the protolophid downward from the metaconid to the carnassial notch, and the degree of development of the talonid cristids. It differs in that it is only three-fourths the size of the M_1 of *P. haydenianus;* its talonid is equal to its trigonid in width, not narrower; and it has three distinct talonid cusps lingual to the hypoconid instead of only one.

There is a very close similarity between this specimen and a lower molar from the early Eocene of Belgium identified by Teilhard de Chardin (1927*a*:21, fig. 18*d*) as an indeterminate miacid. Their trigonids are equal in height, their metaconids are noticeably lower than their protoconids, and their talonids are bordered by a semicircular ridge that has three distinct cusps on the lingual side. There can be little doubt that they were closely related.

TABLE 24

MEASUREMENTS OF CF. MIACIDAE LOWER MOLAR

Specimen	Locality	Trigonid width	Trigonid length	Talonid width (talonid slightly damaged)	Total length
M.N.H.N.-Av 5708 Right..........	Avenay	2.8	2.5	2.6	4.0

DESCRIPTION

(Fig. 12*j, k*)

The trigonid of this specimen is a trifle wider than long. The paraconid is lower than the metaconid, subequal in basal dimensions, and anterior to it. The protoconid is smaller in basal dimensions than the metaconid and subequal in height, its labial height above the talonid is about three-fifths the trigonid length. Located about one-fourth the trigonid width lingual to the labial side of the tooth, the protoconid is lateral to a point about one-third the distance from the metaconid to the paraconid. The apex of the protoconid has been truncated by abrasion but in the unworn condition was probably rounded like those of the metaconid and the paraconid. There is little development of a paralophid or protolophid on the anterior or posterior sides of the shallow prefossid, but there is a low barrier on the lingual side formed by the juxtaposition of the bases of the metaconid and the paraconid, which cannot be properly described as a metacristid. On the central half of the prevallid near the base is developed a prominent anterior cingulum that dips laterally at about 30 degrees from horizontal.

The talonid is three-fifths as long as it is wide. Its length is three-fifths and its width equal to that of the trigonid. The entocristid and postcristid form an unbroken, posterolingually convex ridge from the base of the metaconid to the hypoconid. Completing the enclosure of the deep postfossid, the crista obliqua extends from the hypoconid straight to the trigonid, abutting its base at a point about two-thirds the distance from the metaconid to the protoconid. The hypoconid is the largest cusp in basal dimensions; its distance from the labial edge of the tooth cannot be accurately determined due to breakage. Shortest and smallest in basal dimensions of the talonid cusps, the hypoconulid is nevertheless still well developed and located at what was about the midpoint of the posterior edge of the undamaged tooth. Tallest of the talonid cusps are the entoconid and entoconulid which are subequal in height and basal dimensions and are located near the posterointernal corner of the tooth on the posterior and lingual edges respectively. Except for the slightly truncated tips of the protoconid and hypoconid, no wear facets are evident on the specimen.

COMMENTS

This specimen resembles the M_1 of *Didymictis altidens* from the Wasatchian of Wyoming figured by Gazin (1962, pl. 7, figs. 1 and 2) in the bulbous nature of the paraconid, the development of a prominent anterior cingulum, the prominence of the talonid cusps, the arrangement of the cusps and cristids on the talonid, and the slightly narrower width of the talonid in relation to the trigonid. It differs from *D. altidens* in that it is only half its size, has a well-developed protolophid, a lower and longer trigonid, and a shorter talonid.

The M_2 of *Miacis exiguus* (A.M.N.H. 15717, see Matthew, 1915*d*, fig. 29) from the Wasatchian of Wyoming and this specimen are similar in basal dimensions, relative heights and arrangement of the trigonid cusps, the height of the trigonid above the talonid, the presence of a prominent anterior cingulum, and the arrangement of the cristids on the talonid. They differ in that the M_2 of *M. exiguus* has only a strong hypoconid on the talonid while this specimen has four prominent talonid cusps.

cf. Deltatheridia or Carnivora

TABLE 25

MEASUREMENTS OF CF. OXYAENOIDEA OR MIACIDAE LOWER MOLAR

Specimen	Locality	Trigonid width	Trigonid length	Talonid width	Total length
M.N.H.N.-Av 5773 Left................	Avenay	3.5	4.1	...	...

DESCRIPTION

(Fig. 13*a, b*)

The talonid of this tooth is missing and the metaconid has been broken off near its base. The trigonid is slightly longer than wide. Lower than the protoconid, the paraconid is greater in basal dimensions and anterior to the metaconid. The paraconid points vertically, and its base is directly over the anterior root. Although direct comparison cannot be made due to the destruction of the metaconid, the protoconid was certainly higher and larger in basal dimensions. The protoconid is situated about one-fourth the trigonid width medial to the lateral edge of the tooth and is located labial and slightly anterior to the metaconid. A well-developed paralophid with a deep carnassial notch forms the anterior border of the deep prefossid, and the preserved portion of the protolophid indicates an equally prominent posterior border with a similar carnassiform notch. The lingual side of the prefossid is open because of the wide separation of the paraconid from the meta-

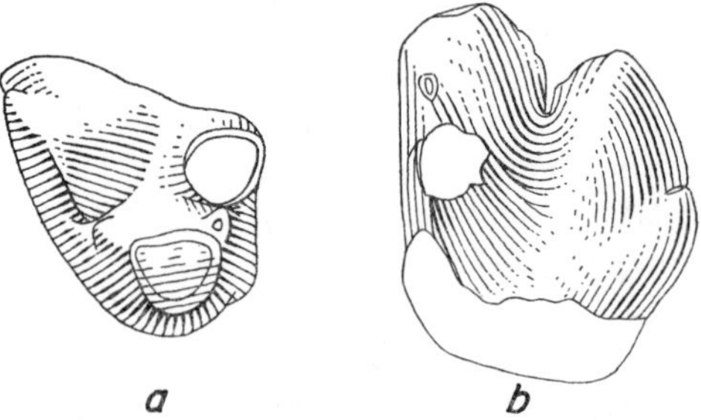

Fig. 13. cf. Deltatheridia or Carnivora, X 6, left trigonid, **M.N.H.N.-Av 5773**, *a,* occlusal view, and *b,* lingual view.

conid. An anterior cingulum that dips downward laterally at an angle of 40 degrees is developed on the labial half of the prevallid. At the base of the para- conid on the anterointernal corner of the tooth is a notch that has nearly closed over to form a small hole with one opening on the lingual surface of the molar and a second on the prevallid. Well-developed wear facets cover the dorsal half of the prevallid, the posterior sides of the protoconid and the preserved fragment of the metaconid.

There is no indication of a point of abutment of the crista obliqua against the posterior face of the trigonid. Quite possibly all evidence of such an abutment was destroyed when the talonid was broken away.

COMMENTS

None of the features preserved on this specimen permit an unequivocal assign- ment of it to either the Carnivora or the Deltatheridia.

Order Condylarthra Cope, 1881*x*
Family Phenacodontidae Cope, 1881*x*
Phenacodus Cope, 1873*d*
Phenacodus teilhardi Simpson, 1929*g*

Lectotype.—Right upper molar illustrated by Teilhard de Chardin, 1927*a*, pl. V, fig. 16.

Phenacodus cf. *P. teilhardi*

A mandible fragment with the P_4-M_3 preserved, and an isolated M_3 both from Mutigny may be provisionally referred to *Phenacodus teilhardi;* the latter speci- men may be thus assigned primarily because of its association in the same quarry with the former. Measurements of these specimens are given in table 26.

DESCRIPTION

(Fig. 14a–c)

The jaw fragment (M.N.H.N.-Louis-250 Mu) consists of the anterior part of the left mandible lateral to the symphysis, and the right mandible complete except for most of the ascending ramus. The labial walls of the incisor alveoli and the partitions between these alveoli are missing. Only the right P_4-M_3 are present, but the alveoli indicate the following dental formula: I_3-C_1-P_4-M_3.

Five foramina are present on the external and ventral surfaces of the right mandible. The first two are paired and located below the alveolus of the canine 3 mm lateral to the midline symphysis. The third is 6 mm below the point where the alveoli of the P_1 and P_2 meet. The fourth is 10 mm ventral to alveolus for the anterior root of the P_4. The fifth is 12 mm ventral to the alveolus of the anterior root of the M_1. On the basis of the position and prominence of the latter three foramina, their function was probably the same as the mental foramen in other, better known mammalian groups.

The masseteric fossa covers the preserved lateral surface of the mandible pos- terior to the M_3 except the ventral-most 10 mm and a narrow ridge along the dorsal

TABLE 26

MEASUREMENTS OF PHENACODUS MANDIBLES

	Almogaver condali[a]	Phenacodus cf. P. teilhardi M.N.H.N.-Louis-250 Mu	Phenacodus cf. P. teilhardi U.C.M.P. 65553	Phenacodus intermedius A.M.N.H. 15761[b]
Length of M_1-M_3	34.3	31.8	...	...
Depth of mandible between between M_2 and M_3	29.6	21.0	...	...
Depth of mandible beneath M_1	26.4	19.5	...	...
Depth of mandible between P_2 and P_3	24.8	17.8	...	21.0
Greatest mandibular width	15.5	12.9	...	...
P_1 Length of alveolus	...	5.2	...	...
Width of alveolus	...	4.0	...	...
P_2 Length of alveoli	7.5	7.5	...	...
Length of anterior alveolus	...	3.3	...	...
Width of anterior alveolus	...	4.1	...	...
Length of posterior alveolus	...	3.5	...	...
Width of posterior alveolus	...	3.9	...	...
P_3 Length of alveoli	...	8.7	...	...
Length of anterior alveolus	...	3.7	...	...
Width of anterior alveolus	...	4.4	...	...
Length of posterior alveolus	...	4.0	...	...
Width of posterior alveolus	...	5.0	...	...
Length	9.0	...	...	...
Width of talonid	5.5	...	...	...
P_4 Length	12.1	9.4	...	10.5
Width of trigonid	8.5	7.5	...	7.3
Width of talonid	8.9	7.1	...	7.1
M_1 Length	11.6	9.8	...	10.8
Width of trigonid	10.4	9.4	...	9.2
Width of talonid	10.6	8.8	...	9.2
M_2 Length	11.4	10.0	...	11.5
Width of trigonid	11.4	10.1	...	10.5
Width of talonid	10.4	9.3	...	9.8
M_3 Length	11.5	10.4	11.3	11.4
Width of trigonid	9.8	8.1	8.5	8.6
Width of talonid	8.2	7.1	6.9	6.9

[a] Data from Crusafont and Villalta, 1954b:69.
[b] All measurements on the left mandible and teeth except M_3.

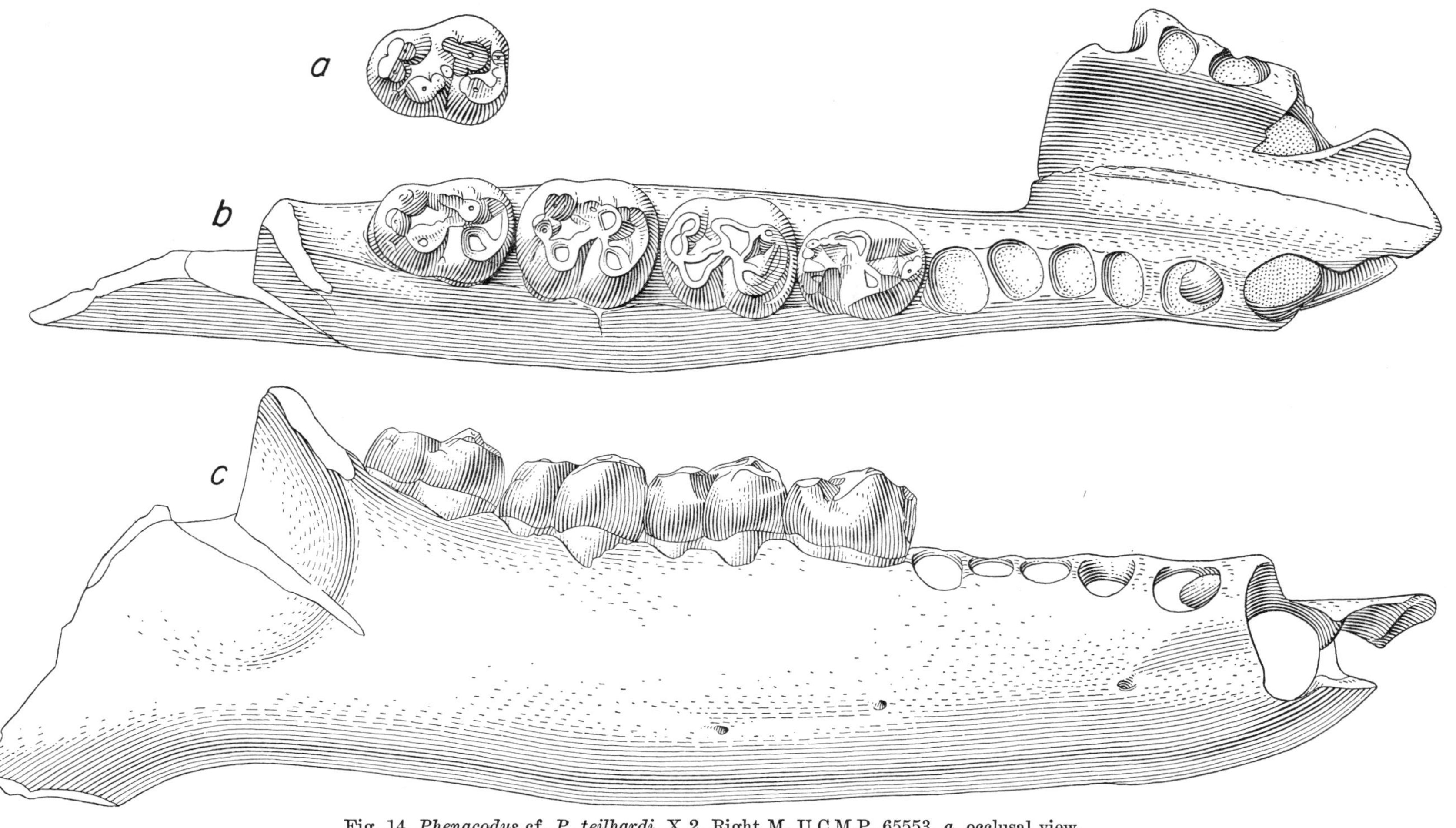

Fig. 14. *Phenacodus* cf. *P. teilhardi*, X 2. Right M_3 U.C.M.P. 65553, *a*, occlusal view.
Right mandible, M.N.H.N.-Louis-250 Mu, *b*, occlusal view, and *c*, labial view.

edge. Beginning about 22 mm below and 8 mm posterior to the M_3, the ventral border is deflected downwards posteriorly. This indicates a prominent angle on the mandible as is known on *Phenacodus intermedius* (A.M.N.H. 15764).

On the medial surface of the mandible there are no foramina. On the posterior edge of the midline symphysis of the jaw is a deep pit that was the site where the geniohyoideus muscle inserted. This pit is bounded ventrally by a sharp mental spine. Posteriorly, this spine diverges from the ventral border of the mandible and becomes less distinct as it extends posteriorly for about 10 mm. From a point 15 mm below and 13 mm posterior to the M_3, the pterygoid fossa covers all that remains of the medial surface of the mandible posterior to this position except for two strips about 5 mm wide which run along the ventral border and along the broken dorsal edge.

I_1–P_3—Alveolar dimensions of I_1 imply that it was a tiny tooth with a root 5.2 mm long and at the base of the crown was 2.3 mm in diameter. Root length and basal crown diameter of the I_2, 6.7 mm and 3.4 mm respectively, indicate a more prominent tooth than either of the other lower incisors. Preserved only on the left side, the tiny I_3 alveolus has a root length of 3.0 mm and a basal crown diameter of 1.7 mm. The base of the canine is circular in cross section with a diameter of 6.7 mm. The alveolar axis indicates that this tooth pointed dorsally and antero-externally and was crescentic with the anterior side convex. The single alveolus of the P_1 is elliptical in outline with the major axis 5.2 mm in length and directed anteroposteriorly. The minor axis is 4.1 mm in length. Within the alveolus is an enigmatic structure which may be a sliver of tooth root that became loose and was subsequently pushed over on its anterior side. The anterior alveolus of the P_2 is also elliptical in outline, but the major axis trends mediolaterally. Lengths of the major and minor axes of this alveolus are 3.5 mm and 3.1 mm respectively. The posterior alveolus is rectangular in shape being 3.5 mm mediolaterally and 3.4 mm anteroposteriorly. Both alveoli of the P_3 are elliptical in outline except where they are jointly straight and contiguous with one another. Maximum dimensions of the anterior alveolus are 3.6 mm anteroposteriorly and 4.3 mm mediolaterally. The same measurements for the posterior alveolus are 3.8 mm and 4.8 mm respectively.

All the teeth preserved are approximately the same height. Weak horizontal striations are present on the vertical surfaces of the P_4 and M_3; similar striations may be seen on the M_2 but are less marked. On the M_1, these striations are barely discernible.

P_4—The P_4 is molariform and slightly narrower and less bulbous than the M_3. There is an appression fossette on the anterior edge of the tooth; the posterior surface cannot be observed. A slight cingulum is developed on the anterolabial corner of the hypoconid base. The trigonid is longer and wider than the talonid. In order of decreasing height, the following trigonid cusps are present: metaconid, protoconid, labial paraconid cusp, and lingual paraconid cusp. The metaconid is posterior and lingual to the protoconid and joined to it by a centrally depressed protolophid. The protoconid has an anterior swelling that meets a posterolabial swelling from the labial paraconid cusp, which is anterior and somewhat lingual to the protoconid. The lingual paraconid cusp is a distinct, well-developed cusp which meets the metacristid (directed anteriorly from the metaconid) and the paralophid (directed lingually from the labial paraconid cusp) to enclose the prominent prefossid.

Only two cusps are present on the talonid: the metastylid and the lower hypoconid. The metastylid is just behind the metaconid on a posteriorly directed entocristid which nearly closes off the lingual side of the postfossid by extending unbroken to the proximity of the postcristid. Occupying the posterolabial corner of the talonid, the hypoconid is separated from the protoconid by a deep V-shaped groove which extends inward to the midline of the tooth where it is separated from the postfossid by the crista obliqua. The crista obliqua extends from the anteromedial corner of the hypoconid to the posterolabial corner of the metaconid and is centrally depressed. The postcristid forms the posterior edge of the postfossid and has two minute swellings upon it in the proper positions for the hypoconulid and entoconid but these cusps can hardly be said to be present on that basis.

M_1—The M_1 is a bulbous tooth with a trigonid that is slightly wider but equal in length to the talonid. A weak cingulum extends along the anterior margin of the tooth from the paraconid to the anterolabial corner of the protoconid. There are two labial cingula developed adjacent to the hypoconid. The first is on the posterior side of the hypoconid. The second is a weak one located near the bottom of the V-shaped groove that separates the hypoconid from the protoconid. Subequal in height, the paraconid and metaconid are taller than the protoconid. The paralophid begins at the anterior corner of the paraconid and extends laterally for about three-fourths its length and there swings posterolaterally to unite with the anterolingual edge of the protoconid. The metaconid is posterolingual to the protoconid and joined to it by a short protolophid. The paraconid is likewise linked to the metaconid by a short anteroposteriorly directed metacristid. The prefossid is shallow and completely enclosed by the surrounding cusps, lophs, and crista.

The following cusps are present on the talonid in order of decreasing height: metastylid, hypoconid, entoconid, and hypoconulid. All these cusps are nearly equal in height and all are lower than the protoconid, the lowest trigonid cusp. The metastylid is posterior and adjacent to the metaconid and scarcely differentiated from it. Unlike the P_4, there is only the slightest indication of an entocristid extending posteriorly from the metastylid towards the entoconid. Hence, the postfossid is somewhat more open on the lingual side than it is on the P_4. The hypoconid is a prominent cusp posterior to the protoconid and posterolabial to the metaconid. The protoconid and hypoconid are separated by a broad V-shaped groove that extends inward toward the crista obliqua and is not as deep as the similar structure on P_4. The crista obliqua is positioned as it is on the P_4 but is wider and lacks the depression in the middle, a difference probably owing to greater wear on the M_1. The hypoconulid is located at the posterior edge of the tooth and is posterolingual to the hypoconid with which it is contiguous. The entoconid is adjacent to the anterolingual corner of the hypoconulid. Both of these cusps are somewhat smaller in basal dimensions than the hypoconid but are still quite prominent.

M_2—The M_2 differs from the M_1 in the following characters: the postfossid is deeper; the paralophid is less distinct; the metastylid is distinct from the metaconid; a poorly defined, short entocristid extends anteromedially from the antero-medial edge of the entoconid; and the anterior cingulum and the cingulum anterior to the hypoconid are absent.

M_3—The M_3 is a narrower and less bulbous tooth than the anterior molars. The trigonid is wider and shorter than the talonid. The cusps, lophs, and crista on the trigonid are the same as the M_1 with the following exceptions: the metaconid is nearer to the protoconid, hence the protolophid is shorter; and similarly, the meta-conid and paraconid are closer together and thus the metacristid is shorter. The prefossid is shallower than on the M_1. The talonid is markedly narrower relative to its length than on the anterior lower molars. Posteriorly directed from the meta-stylid, the entocristid terminates in a weak cuspule that forms the anterior edge of the lingual opening of the postfossid. The entocristid was not found on the anterior molars but its presence was noted on the P_4. On the anterolingual side of the entoconid is another cuspule that forms the posterior edge of the postfossid lingual opening. As on the M_2 but unlike the M_1, there is no anterior cingulum. A distinct ectostylid is present between the bases of the protoconid and hypoconid.

The isolated right M_3 (U.C.M.P. 65553) of *Phenacodus* cf. *P. teilhardi* (also from Mutigny) shows some differences from the M_3 in the jaw just described that are the result of wear and several more that may be attributed to genetic differ-ences between the individuals that possessed these teeth. The trigonid is much wider than the talonid instead of being only slightly wider. Though the tooth is more heavily worn, its prefossid is slightly deeper and its paralophid more promi-nent. The paraconid is larger in basal dimensions and separated from the meta-conid by a distinct narrow groove rather than in juxtaposition with it. The lingual opening of the postfossid is much larger due to the absence of mammelons and an ectocristid between the entoconid and metastylid. The hypoconulid is somewhat larger in basal dimensions and slightly higher, not lower, than the entoconid. There is no trace of a distinct ectostylid between the protoconid and hypoconid.

COMMENTS

Six isolated teeth from three European countries have been previously referred to *Phenacodus teilhardi*. These include three molars from Orsmael, Belgium, described as the types of *P. europaeus* Teilhard de Chardin, 1927a. *P. teilhardi* Simpson, 1929g, was proposed as a replacement for *P. europaeus* Teilhard de Chardin for that name was a junior homonym of *P. europaeus* Rütimeyer, 1888a, a species that was later found to be an artiodactyl and renamed. Two of these molars including the lectotype are uppers (M^1 or M^2) and the third is a lower M_1 or M_2. Referral of the other two teeth mentioned above from Orsmael, particularly the lower molar, to the same species as the lectotype seems highly probable. Other specimens include two upper molars (both M^3) from Epernay, France, referred tentatively by Teilhard de Chardin (1922a:57) to the same species as the lower molar from Orsmael and therefore to *P. teilhardi* and an isolated P_4 from Spain that was also tentatively referred (Crusafont, 1957c). Of all this material, only the lower molar from Orsmael and the P_4 from Spain are of any value for comparison with the Mutigny *P.* cf. *P. teilhardi* specimens.

The M_1 and M_2 of *Phenacodus* cf. *P. teilhardi* from Mutigny are at least 25 percent larger in every dimension than the isolated lower molar of *P. teilhardi* from Orsmael. Their trigonids are lower relative to their talonids than on the isolated lower molar of *P. teilhardi* from Orsmael. The Orsmael lower molar has a broad cingulum posterior to the hypoconid that is absent from the lower molars from Mutigny. The most apparent differences are the pronounced crenulation and doubled cusps of the isolated lower molar of *P. teilhardi* that are absent from the lower molars from Mutigny.

Only a minor difference in size and the presence of a weak basal cingulum on the P_4 tentatively referred to *Phenacodus teilhardi* by Crusafont, 1957c, differentiates it from the P_4 of *P.* cf. *P. teilhardi* from Mutigny.

It is difficult, if not impossible, to identify isolated teeth of *Phenacodus* to the specific level. On the basis of the Belgian material referred to *P. teilhardi,* that species can be separated from *P. intermedius* only geographically as Teilhard recognized (1927a:22) and morphologically in that the enamel of the Belgian material is more complicated, a character which even Cope, not commonly thought of as a lumper, recognized as being a variable trait within a given species of *Phenacodus* (1884o:440).

The Mutigny mandible is the first specimen that provides some morphological basis for separating *Phenacodus teilhardi* from the other species of *Phenacodus* and *Tetraclaenodon*. The P_4 is distinct from all North American species and from *P. villata* Crusafont, 1956b, from the Eocene of Spain in the presence of a metacrista closing off the lingual side of the prefossid and an enlarged internal paraconid cusp. It differs from *Almogaver condali* in having a markedly shallower mandible.

The Mutigny *Phenacodus* cf. *P. teilhardi* material is very similar to the type specimen of *P. intermedius* (A.M.N.H. 15761, see Granger, 1915a, fig. 5) from the Wasatchian of North America. The nature of the type specimen restricts comparisons primarily to dental features but it should be noted in passing that the depths and breadths of the mandibles are very similar.

P_4—The greatest differences in dental morphology are to be found in the structure of the P_4. The Mutigny P_4 has a small lingual paraconid cusp as well as a labial paraconid cusp while the type specimen of *Phenacodus intermedius* has only the latter cusp. However, a referred specimen of *P. intermedius* also from the Big Horn Basin (A.M.N.H. 15764) has both cusps. The Mutigny P_4 has a strong metacristid extending from the metaconid to the paraconid that forms the lingual wall of the prefossid. On *P. intermedius,* as on all North American specimens of *Phenacodus,* the metacristid is absent on P_4 leaving the prefossid open lingually. The postcristid and entocristid are more developed on the Mutigny P_4 leaving the postfossid more enclosed than on the type specimen of *P. intermedius*. However the referred specimen of *P. intermedius* has these structures well developed on the P_4.

M_1—The small differences that are present between the M_1 of the Mutigny jaw and the type specimen of *P. intermedius* can be largely attributed to the greater wear on the former specimen. The only possible exceptions are the presence of a cingulum along the anterior border of the tooth in front of the paraconid and the wider cingula on the labial half of the anterior and posterior borders of the type specimen of *P. intermedius*.

M₂—The same statements made for the comparison of the M_1 can be made for the M_2 with the exception of a slightly greater difference due to the lack of an anterolabial cingulum on the Mutigny M_2.

M₃—The two Mutigny M_3 specimens have only two characters in common that are different from the type specimen of *Phenacodus intermedius:* they lack the development of a cingulum along the anterior edge of the tooth and have a narrower marginal cingulum between the hypoconid and hypoconulid. The Mutigny M_3 preserved in the jaw fragment differs from that of the type specimen of *P. intermedius* in the near closing of the lingual side of the postfossid by the extension of the entocristid from the metastylid to the vicinity of the entoconid. The isolated Mutigny M_3 differs from that of *P. intermedius* in the presence of a larger paraconid and the complete lack of an entocristid extending from the metastylid halfway to the entoconid.

Comparison between a cast of the type specimen of *Tetraclaenodon transitus* (U.C.M.P. 34762, see Dorr, 1958a, pl. 2, figs. 1–4) from the Torrejonian of Wyoming and the Mutigny *Phenacodus* cf. *P. teilhardi* must be restricted to comparisons of the P_4–M_2. The individual cheek teeth of the Mutigny jaw are smaller with the cusps more inflated, and the P_4 talonid is markedly narrower relative to the M_1 trigonid.

P₄—As in the comparison with the type specimen *Phenacodus intermedius,* the greatest degree of morphological dissimilarity is to be found on the P_4. The talonid of the Mutigny P_4 is slightly narrower than the trigonid rather than markedly wider. There is a slight expansion of the paralophid in the region of the labial paraconid cusp of *Tetraclaenodon transitus* but no lingual paraconid cusp. The metacristid and entocristid are absent from *T. transitus* leaving the prefossid and postfossid open lingually in contrast to the P_4 from Mutigny. A short broad labial cingulum that is developed on the anterior side of the hypoconid on the P_4 of *T. transitus* is absent from the P_4 on the specimen from Mutigny.

M₁—The M_1 differs in that *Tetraclaenodon transitus* lacks any trace of a metastylid and has an entocristid extending a short distance anteriorly from the entoconid. *T. transitus* has a strong labial cingulum developed between the protoconid and hypoconid and a shorter but equally strong cingulum posterior to the hypoconid. The M_1 from Mutigny completely lacks the former cingulum and the cingulum posterior to the hypoconid is only weakly developed. A weak cingulum is developed on the anteroexternal corner of the Mutigny molar that is not present on *T. transitus.*

M₂—The M_2 of *Tetraclaenodon transitus* does not show quite as great a degree of difference from the corresponding tooth on the Mutigny specimen as the M_1, for the labial cingulum between the protoconid and hypoconid is much smaller and restricted to the posterior side of the former cusp. A weak cingulum is present on the anteroexternal corner of the M_2 of *T. transitus* and absent on the M_2 from Mutigny. Except for these two exceptions, the M_2 comparisons between *T. transitus* and the *Phenacodus* cf. *P. teilhardi* material from Mutigny are the same as the M_1 comparisons.

The type and only known specimen of *Almogaver condali* is a right mandible with P_4–M_3 present from the medial Eocene (Lutetian) of Spain. The overall dental and mandibular morphology of *A. condali* and the *Phenacodus* cf. *P. teilhardi* material from Mutigny is quite similar, most of the major differences being of a quantitative rather than qualitative nature. Many of these quantitative differences are apparent in table 26. The mandible of *P.* cf. *P. teilhardi* is longer, narrower mediolaterally, and shallower dorsoventrally. Also, its greatest mediolateral width is beneath the M_2 rather than beneath the contact between the M_1 and M_2. Though the total length of the molar-premolar series is nearly the same, the length of the premolar series is greater than the molar series; the reverse is true on *A. condali*.

On *Phenacodus* cf. *P. teilhardi* there is a second prominent foramen on the external side of the mandible beneath the point of contact between the P_1 and P_2, which was not reported on *Almogaver condali*. Both mandibles display a well-developed mental spine. The preserved part of the ascending ramus of both jaws indicates the ascending rami were almost vertical.

P_4—*Phenacodus* cf. *P. teilhardi* has a protoconid that is subequal to the metaconid in basal dimensions rather than being larger as in *Almogaver condali*. On both, the metaconid is higher than the protoconid and posterolingual to it. The labial paraconid cusp is prominent on both specimens. The metacristid of *A. condali* seems [on the basis of the clearest illustration of P_4 (Crusafont and Villata, 1954*b*, pl. 14)] to be as prominent as that of *P.* cf. *P. teilhardi*. On these two specimens the metastylid together with the other trigonid cusps and lophs completely encloses the prefossid forming a deep basin, a condition unique within the *Tetraclaenodon-Phenacodus* complex where the prefossid is usually open on the lingual side. The morphology of the talonid is the same on both specimens. In particular, the degree of strength and the position of the hypoconid, and the development of the postcristid and crista obliqua are the same. Crusafont (pers. comm., 1967), who has made direct comparison between a cast of *P.* cf. *P. teilhardi* and the type of *A. condali*, notes that the P_4 talonid of the former is narrower than that of the latter.

M_1—*Phenacodus* cf. *P. teilhardi* has a trigonid that is slightly wider rather than somewhat narrower than the talonid. The positions and relative heights of the trigonid cusps are the same. A metastylid was not reported on *Almogaver condali*, a fact that is readily understandable because it survives as a mere remnant obscured by wear on *P.* cf. *P. teilhardi*. Based on the description, it seems that the crista obliqua of *A. condali* is weaker, but comparison between *P.* cf. *P. teilhardi* and the best figure of the M_1 of *A. condali* (Crusafont and Villata, 1954*b*, pl. 14) does not seem to bear this out. The hypoconid of the Mutigny specimen is slightly lower than the entoconid rather than higher, but the basal dimensions and positions of all the talonid cusps are the same. There is no minute accessory cusp on the anterior side of the entoconid as in *A. condali*. On both there is a deep groove between the entoconid and metastylid that forms the lingual opening of the postfossid. A weak external cingulum is common to the Mutigny specimen and *A. condali*.

M_2—As on *Almogaver condali*, the trigonid of the M_2 of *Phenacodus* cf. *P. teilhardi* is wider than the talonid. On *P.* cf. *P. teilhardi* the M_2 is longer than the M_1 rather than shorter as on *A. condali*. The paraconid is as prominent on the M_2

as on the M₁ of *P.* cf. *P. teilhardi* in contrast to *A. condali* on which it is weaker. Crusafont (pers. comm., 1967) reports that the hypoconulid is stronger on *A. condali* than on *P.* cf. *P. teilhardi*. Except for the few differences listed immediately above, the comparisons between the M₂ of *P.* cf. *P. teilhardi* and *A. condali* were the same as those given for the M₁.

M₃—The M₃ of *Phenacodus* cf. *P. teilhardi* is slightly longer relative to its maximum width but is otherwise similar in outline to the M₃ of *Almogaver condali*. The prefossid is narrower and just as deep as on the preceding molars, not wider and shallower as on *A. condali*. A weak metastylid is present on all three specimens of M₃. The M₃ of *A. condali* has a weak entoconid and a hypoconulid that is markedly stronger than the one on the M₂ in contrast to the M₃ of *P.* cf. *P. teilhardi* on which the entoconid is well developed and the hypoconulid is no stronger than on the M₂. One M₃ of *P.* cf. *P. teilhardi* (M.N.H.N.–Louis–250 Mu) has a mammelon on the anteromedial edge of the entoconid that is not found on the other M₃ of this species known from Mutigny (U.C.M.P. 65553) nor on the M₃ of *A. condali*. In contrast to *A. condali*, the crista obliqua of the *P.* cf. *P. teilhardi* M₃ is no more marked in its development than on the M₁ or M₂, and there is no trace of an internal cingulum.

Thus the only major qualitative difference between *Almogaver condali* and the *Phenacodus* cf. *P. teilhardi* appears to be the marked difference in the form of the M₃.

The quantitative comparisons are listed in table 26. The comparison of the lengths and widths of the teeth revealed some differences slightly greater in proportional range than were obtained by Simpson (1937*d*:18), in a single population of *Phenacodus primaevus*. However, as Simpson (1947*a*) pointed out, the use of range as a statistic for comparison between two samples, especially small ones, is apt to be misleading for the real limits of variation will often be significantly greater. The differences in mandibular depths of the two specimens are less than may be seen between individuals of the somewhat similar modern species *Tayassu tajacu* (the peccary).

The great similarity in dental morphology between *Almogaver condali* and the *Phenacodus* cf. *P. teilhardi* mandible from Mutigny suggests a close phylogenetic relationship. Their P₄ structure is unique among the specimens referred to the *Phenacodus-Tetraclaenodon* complex in the development of a doubled paraconid with the labial cusp stronger. Another feature of this tooth exclusive to both

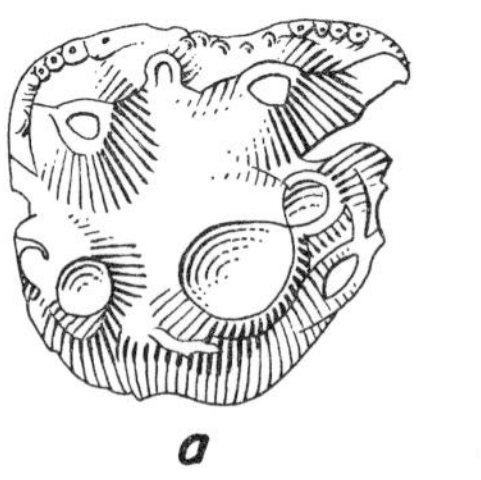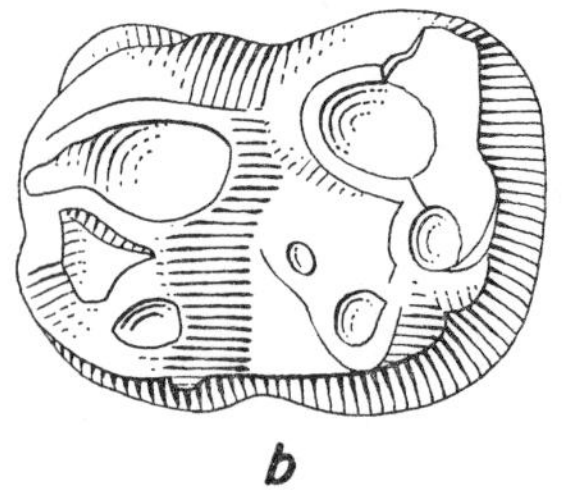

a   *b*

Fig. 15. *Phenacodus* sp., X 3, *a*. Right DP⁴, M.N.H.N.-Louis-196 Mu, occlusal view. *b*. Left M₁ or M₂, M.N.H.N.-Louis-83 Gr, occlusal view.

specimens (if it is present on *A. condali*) is the development of a strong meta-cristid which closes off the lingual side of the prefossid. Their relative geochrono-logic age indicates that *P.* cf. *P. teilhardi* is ancestral to *A. condali*.

Phenacodus sp.

Two isolated teeth, a right DP⁴ and a left M₁ or M₂, may be tentatively referred to *Phenacodus*.

TABLE 27

MEASUREMENTS OF *PHENACODUS* SP. DP⁴

Specimen	Locality	Maximum length	Maximum width
M.N.H.N.-Louis-196 Mu Right	Mutigny	9.0	8.8

DESCRIPTION

(Fig. 15*a*)

DP⁴—The crown of this specimen is complete except for a small triangular chip missing from the region anterolingual to the paracone. Its outline in occlusal view is a trapezoid with the labial edge forming the base. Prominent appression fossettes are present on the anterior and posterior borders of the tooth. The metacone is posterior and slightly lingual to the paracone. Both cusps are equal in height and basal dimensions. The mesostyle is equidistant from them and situated about two-thirds as far from the labial border of the tooth as the paracone is. The meso-style is lower than the paracone and metacone but higher than the remaining cusps. As is the usual case, the mesostyle is markedly smaller in basal dimensions than the paracone and metacone. A weak paracrista extends from the anterior border of the tooth to the paracone. Linking the paracone, mesostyle, and meta-cone is an equally weak centrocrista. The miniscule metastyle is joined to the metacone by a weak metacrista.

The lingual area of the tooth has been much more abraded than the labial region. Consequently the protocone, protoconule, and hypocone are quite low and sub-equal in height. The basal dimensions of the protocone are equal to those of the paracone and metacone; presumably when unworn, the three cusps were equal in height. The protocone is somewhat less anteriorly placed on the tooth than the paracone. The protoconule is equidistant from the paracone and protocone and slightly anterior to the point midway between them. Its basal dimensions are the same as the mesostyle, and presumably it was as high as that cusp when unworn. Posterior to the protocone, the hypocone is slightly more posteriorly placed on the tooth than the metacone. Its basal dimensions are intermediate between the larger and smaller cusps and so too, presumably, was its unabraded height. No sign of a metaconule is present. Slight lingual elongations of the paracone and metacone might be described as a protoloph and metaloph respectively, but other than these faint traces, the two lophs are absent.

Marginal cingula are developed all along the undamaged anterior, posterior, and labial edges of the tooth except where interrupted by the paracrista and bases of the metastyle, mesostyle, and paracone. Lingual to the posterior part of the base of the protocone is a short cingulum. Where unworn, the marginal cingula are crenulated.

COMMENTS

Robert West (pers. comm., 1969) has pointed out several features of this tooth that support its identification as a *Phenacodus* DP^4. The prominent parastyle gives the tooth a trapezoidal outline that strongly contrasts with the rectangular shape of the *Phenacodus* upper molar but is characteristic of the DP^4. Reduction of the hypocone is also characteristic of the *Phenacodus* DP^4. Paraconule position slightly in advance of a line between the protocone and paracone, and the metaconule on the line between the hypocone and metacone are characteristic of *Phenacodus* upper cheek teeth.

TABLE 28

MEASUREMENTS OF PHENACODUS SP. M_1 OR M_2

Specimen	Locality	Trigonid width	Trigonid length	Talonid width	Total length
M.N.H.N.-Louis-83 Gr Left..............	Grauves	9.3	...	9.0	10.4

DESCRIPTION

(Fig. 15*b*)

M_1 or M_2—On the trigonid there are two cusps, the protoconid and metaconid, the latter being larger in basal dimensions and somewhat more posterior on the tooth than the former. The comparative basal dimensions of cusps on teeth as heavily worn as this one are apt to be a highly misleading estimate of the relative sizes of the unworn cusps because what must be compared are the dimensions of dentine lakes, which are determined in part by the degree of wear the tooth has undergone as well as by the original sizes of the cusps. The bases of the protoconid and metaconid merge; because the state of wear is advanced, it is impossible to determine whether a protolophid ever joined them. The metacristid extends forward from the metaconid to the anterior border of the tooth. The paralophid begins at the anterior end of the metacristid and extends laterally along the anterior margin of the tooth swinging to the posterolateral direction for the last half of its length to join the protoconid. The shallow prefossid is completely enclosed by the surrounding cusps and lophs. No trace could be seen of either a paraconid or parastylid, but such features could easily have been obliterated by abrasion if they were small. On a cast of the mandible of *Phenacodus* cf. *P. teilhardi* from Mutigny described above, the M_1 and M_2 were artifically worn down. One of the results was that the swelling at the base of the paraconid on the M_1 after slight abrasion appeared to be a broad metacristid much as in this isolated lower molar.

The talonid is wider and only slightly lower than the trigonid. The three talonid cusps are equal in height. The hypoconid is markedly smaller in basal dimensions and more anteriorly placed on the tooth than the entoconid. The hypoconulid is posterior to both of these cusps and slightly larger than the hypoconid. The crista obliqua is not prominent, being a broad, poorly defined ridge dividing the post-fossid from the V-shaped valley between the hypoconid and the protoconid. Lack of an entocristid between the metaconid and entoconid leaves the lingual side of the postfossid open. No marginal cingula are present on the tooth.

COMMENTS

The majority of differences between this specimen and the M_1 of the type specimen of *Phenacodus intermedius* (A.M.N.H. no. 15761, see Granger, 1915a, fig. 5) can be attributed to the greater degree of abrasion that the Grauves specimen has undergone. The differences in this category are the presence of a paraconid, and the greater cusp heights on *P. intermedius*. The only differences that might not fall into this category are the presence of a narrow cingulum along the entire anterior border and a short but wide cingulum labial to the hypoconulid and posterior to the hypoconid on *P. intermedius*. Neither of these cingula are present on the Grauves specimen. The dimensions of the two teeth are nearly identical, the disposition of the cusps and lophs are the same, and the degree of inflation of the cusps is equal. If isolated teeth of *Phenacodus* and *Tetraclaenodon* could be assigned confidently to species, there would be reason to refer the Grauves specimen to *P. intermedius*.

However, comparison with the lower molars of *Tetraclaenodon transitus* does not reveal a markedly greater degree of difference. The majority of differences between the M_1 of *T. transitus* and the Grauves tooth may also be attributed to abrasion. In these two specimens, the degree of abrasion is about the same, but the pattern of wear is different; the M_1 of *T. transitus* has the greatest wear on the labial cusps and the Grauves molar has the greatest wear on the lingual ones. Quite predictably, there appears to be a marked difference in the basal dimensions of corresponding cusps of the two teeth. Though this disparity may indicate a basic dissimilarity in grinding mode, it seems more likely to have been a function of individual variation, for a slight difference in the angle of occlusion between the upper and lower molar surfaces in grinding teeth such as these could readily account for the degree of diversity seen in this instance. The M_1 of *T. transitus* has a prominent labial cingulum between the protoconid and hypoconid and another short but equally wide cingulum labial to the hypoconulid and posterior of the hypoconid. Neither of these cingula is present on the Grauves specimen. It should be noted, however, that on the M_2 of *T. transitus* the former cingulum is present but is much reduced.

There are no significant differences between the anterior lower molars of *P. intermedius*, *P. teilhardi*, and *Almogaver condali*, and thus there is no point in comparing the Grauves lower molar with the latter two species.

The similarity of this isolated tooth to the anterior lower molars of *Tetraclaenodon transitus* makes the generic assignment to *Phenacodus* doubtful. The somewhat greater similarity to the type specimen of *P. intermedius* (though the difference in degree of similarity is small and the characters that show the greater affinity are of doubtful validity) together with the known stratigraphic ranges of *Phenacodus* and *Tetraclaenodon* make assignment to the former genus more probable.

Order incertae sedis
Family Paroxyclaenidae Weitzel, 1933*d*
cf. *Paroxyclaenus* Teilhard de Chardin, 1922*a*

TABLE 29

MEASUREMENTS OF CF. PAROXYCLAENUS M[1]

Specimen	Locality	Maximum length	Maximum width
M.N.H.N.-Av 5714 Left....	Avenay	3.4	4.9

DESCRIPTION

(Fig. 16*a*)

The general outline of this tooth in occlusal view is a nearly rectangular trapezoid with the labial side forming the base. The labial edge has only a faint indication of an ectoflexus, the anterior and posterior sides are straight, and the lingual edge is semicircular. The paracone is higher and larger in basal dimensions than the metacone and located anterior to it. The height of the paracone above the contact between the root and the crown enamel is two-thirds the length of the tooth. The tip of the paracone has been flattened by abrasion but was probably rounded like that of the metacone when unworn. The protocone is equal to the paracone in height and equal to the metacone in basal dimensions. The tip of the protocone is rounded and broader than that of the metacone. Although there is no precingulum or postcingulum, the protocone is surrounded on the anterior, posterior, and lingual sides by a broad expansion of the tooth. The prominent paraconule is more

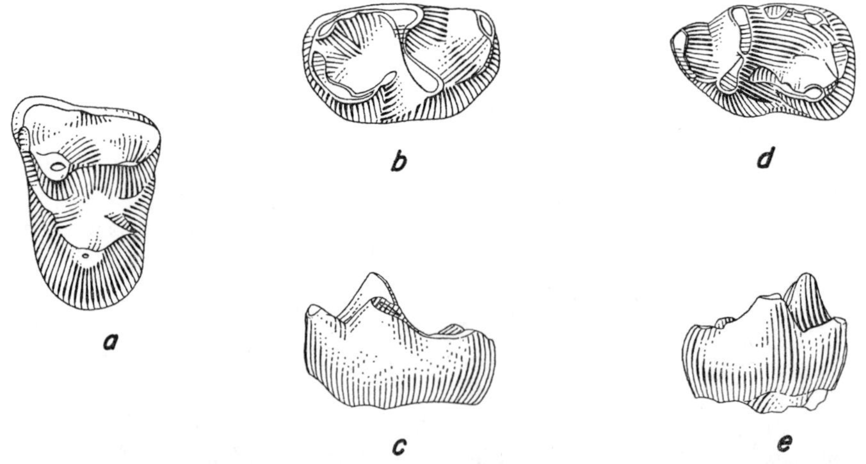

Fig. 16. cf. *Paroxyclaenus*, X 6. Left M[1], M.N.H.N.-Av 5714, *a*, occlusal view. Right M[1], M.N.H.N.-Av 5909, *b*, occlusal view, and *c*, lingual view. Left M[2], M.N.H.N.-Mu 5947, *d*, occlusal view, and *e*, lingual view.

lingually placed on the tooth than the equally prominent metaconule. Linking the paraconule to the protocone is a well-developed preprotocrista. Adjacent to the metaconule on the strong postprotocrista is a well-developed carnassiform notch. A groove extends from this notch almost to the middle of the protofossa. A well-developed paracingulum extends along the anterior edge of the tooth from the paraconule to a point midway between the paracone and the labial edge of the tooth where it is confluent with the paracrista. The paracrista extends from the anterolabial corner of the paracone out to the anterolabial corner of the tooth where it meets the ectocingulum. There is no metacingulum and only an indistinct metacrista, which extends from the posterolabial corner of the metacone to the posterolabial corner of the tooth where it meets the ectocingulum. The centrocrista is low but present between the paracone and metacone. The only cusp on the broad stylar shelf is a faint mesostyle on the ectocingulum. The ectocingulum is continuous along the labial edge of the tooth from one end to the other.

COMMENTS

The general appearance of this tooth from Avenay resembles that of a cast of the M^1 on the type of *Paroxyclaenus lemuroides* Teilhard de Chardin, 1922a, from the ?late Eocene of France (A.M.N.H. 55956, see Teilhard de Chardin, 1922a, pl. 4, fig. 15). They are alike in outline, position, and morphology of the cusps, absence of stylar cusps, presence of a low but distinct centrocrista between the well-separated paracone and metacone, and the absence of a lingual cingulum and hypocone. On *P. lemuroides* the M^1 metastylar region is slightly wider than the parastylar region, but on this specimen the reverse is true.

A cast of the M^1 on the type specimen of *Russellites simplicidens* Van Valen, 1965a (A.M.N.H. 45768, see Van Valen, 1965a, figs. 2–3) from the medial Eocene (Lutetian) of Egerkingen, Switzerland, differs from the Avenay M^1 in having a much narrower stylar shelf, weaker paraconule and metaconule, and a paracone much larger in basal dimensions and height relative to the metacone.

In contrast to this M^1, figures and description of the upper molar of *Dulcidon gandaensis* (Dehm and zu Oettingen-Spielberg, 1958) from the medial Eocene of Pakistan in Dehm and zu Oettingen-Spielberg (1958:19-21; abb. 3; Tafel 1, abb. 6 a–b) indicate that the paraconule is more lingual relative to the metaconule and a postcingulum, precingulum, postparaconule wing, and ectoflexus are all present.

The description and figures of *Kochictis centenii* Kretzoi, 1943a, from the late Oligocene of Hungary given in Kretzoi (1943a) provide enough information to distinguish immediately the upper molars of that species from this M^1. The upper molars of *K. centenii* differ in having a postcingulum, and weaker paraconule and metaconule if these cuspules are present at all.

From the published accounts of the M^1 and M^2 of *Pugiodens mirus* Matthes, 1952, from the medial Eocene (Lutetian) of East Germany in Matthes (1952:230; Tafel xxxvii, fig. 59; Tafel xxxviii, fig. 62), and Van Valen (1965a:390–391; fig. 1) there is little to distinguish them from this M^1. However, the following differences do exist: M^1 and M^2 of *P. mirus* are noticeably wider mediolaterally relative to their length, shorter anteroposteriorly on their lingual side relative to their labial side, and they lack a paracingulum.

From comments about *Kopidodon macrognathus* (Wittich, 1902*a*) from the medial Eocene (Lutetian) of West Germany in Crusafont and Russell (1967) and Weitzel (1933*d*) enough information is available to distinguish the dentition of that species from the M^1 described above. The upper molars of *K. macrognathus* have weaker metaconules and protoconules.

Prominent differences between this M^1 and the M^2 of *Spaniella carezi* Crusafont and Russell, 1967, from the early Eocene (Cuisian) of Spain (see Crusafont and Russell, 1967, fig. 1) are the much stronger paraconule and metaconule, much broader stylar shelf, and smaller size of the former.

TABLE 30

MEASUREMENTS OF CF. PAROXYCLAENUS LOWER MOLARS

Specimens	Locality	Trigonid width	Trigonid length	Talonid width	Total length
M.N.H.N.-Av 5909 Right M_1	Avenay	2.8	2.7	2.7	4.3
M.N.H.N.-Mu 5947 Left M_2	Mutigny	2.3	2.4	2.5	4.0
M.N.H.N.-Mu 6385 Right M_2	Mutigny	2.4	2.5	2.5	4.1

DESCRIPTION

(Fig. 16*b–e*)

M_1—The width of the trigonid is about the same as its length. The paraconid is lower and smaller in basal dimensions than the metaconid and is located anterior to it. The protoconid, the highest trigonid cusp, is intermediate in basal dimensions between the paraconid and metaconid, and is one-fourth the trigonid width lingual to the labial edge of the tooth. It is lateral and slightly anterior to the metaconid. The trigonid cusps are all worn to varying degrees; therefore the original form of their apices cannot be determined. The low protolophid is the strongest of the three trigonid intercusp crests. The paralophid and metacristid are much weaker. There is no indication of an anterior labial cingulum on the trigonid.

The talonid is three-fifths as long as it is wide. It is three-fifths as long as the trigonid and about as wide. The entocristid and postcristid form a continuous, posterolingually convex ridge from the base of the metaconid to the hypoconid. The crista obliqua, somewhat laterally concave, extends from the hypoconid to the base of the trigonid at a point equidistant from the protoconid and metaconid. The hypoconid is the most distinct talonid cusp and is slightly more labially displaced from the tooth midline than the protoconid. The hypoconulid is nearly twice the size of the entoconid but both are mere swellings on the entocristid and postcristid. The postfossid is completely enclosed and deep. The only wear facets are subhorizontal abrasion surfaces on the crests of the cusps and cristids.

M_2—These teeth are about 15 percent smaller than the M_1. A similar size relation exists between the M_1 and M_2 of *Paroxyclaenus lemuroides*. The talonid of these teeth is two-thirds the length of the trigonid rather than three-fifths as on the M_1. Other than these minor differences, the M_2 is the same as the M_1.

COMMENTS

The M_1 and M_2 are similar to the same teeth on *Paroxyclaenus lemuroides*. Damage and the degree of wear apparent on a cast of the lower dentition of *P.*

lemuroides (A.M.N.H. 55956, see Teilhard de Chardin, 1922a, pl. 4, figs. 13, 14) do not permit a complete comparison. However, similarities exist in the general outline of the teeth, height of the trigonid and protoconid relative to the talonid, position and development of the talonid crests, and prominence of the hypoconid. The only clear difference is that the talonid on the M_1 and M_2 of *P. lemuroides* is slightly narrower than the trigonid while on these specimens it is slightly wider.

As figured and described by Kretzoi (1943a), the lower molars of *Kochictis centenii* are immediately separable from this M_1 and M_2 by the labially shifted position of their paraconid relative to their metaconid, and the greater relative length of their talonid compared to their trigonid.

Few differences are evident between the description and figures of the P_4 of *Pugiodens mirus* given in Matthes (1952:233; Tafel XL, figs. 67–71) and the lower molars described above. From the measurements given of the maximum width of the tooth (4 mm) and the length of the trigonid (3 mm), the P_4 trigonid of *P. mirus* is much more expanded anteroposteriorly relative to its width. Also, the P_4 talonid of *P. mirus* is narrower rather than wider than the trigonid.

Kopidodon macrognathus lower molars have a taller trigonid and more reduced talonid and paraconid according to Weitzel (1933d) and Crusafont and Russell (1967).

Distinguishing the M_2 of *Spaniella carezi* (see Crusafont and Russell, 1967, figs. 2, 3) from this M_1 and M_2 are the smaller angle between the prevallid and postvallid, greater width relative to length, and presence of a labial cingulum on the trigonid of the former.

cf. Paroxyclaenidae

TABLE 31
MEASUREMENTS OF CF. PAROXYCLAENIDAE LOWER MOLARS

Specimens	Locality	Trigonid width	Trigonid length	Talonid width	Total length
M.N.H.N.-Av 5898 Left M_1?	Avenay	3.4	...	3.0	...
M.N.H.N.-Av 6759 Right M_1?	Avenay	3.8	1.9	3.6	4.1
M.N.H.N.-Louis-124 Av Left M_2?	Avenay	3.3	2.6	3.2	...
M.N.H.N.-Louis-189 Mu Right M_2?	Mutigny	3.0	2.0	3.0	3.8
M.N.H.N.-Av 5002 Left M_2?	Avenay	3.1	2.2	...	...
M.N.H.N.-Av 5731 Right M_2?	Avenay	3.0	1.7	3.0	3.5
M.N.H.N.-Mu 5942 Right M_2?	Mutigny	2.9	2.1	2.8	3.9
M.N.H.N.-Mu 6168 Left M_2?	Mutigny	3.2	...	3.1	3.7
M.N.H.N.-Mu 6272 Left M_2?	Mutigny	2.9	2.3	2.9	...
M.N.H.N.-Mu 6304 Left M_2?	Mutigny	...	1.7	...	3.5
M.N.H.N.-Louis-23 Py Right M_3?	Pourcy	2.4	1.7	2.1	3.0
M.N.H.N.-Louis-49 Py Right M_3?	Pourcy	2.3	1.7	2.2	3.1
M.N.H.N.-Louis-190 Mu Right M_3?	Mutigny	2.7	1.5	2.7	3.3
M.N.H.N.-Av 5656 Right M_3?	Avenay	3.0	...	3.0	3.0

DESCRIPTION

(Fig. 17a–o)

M_1?–M_3?—The trigonids of this group of molars are anteroposteriorly compressed; their length is only one-half to three-fourths their width. The paraconid

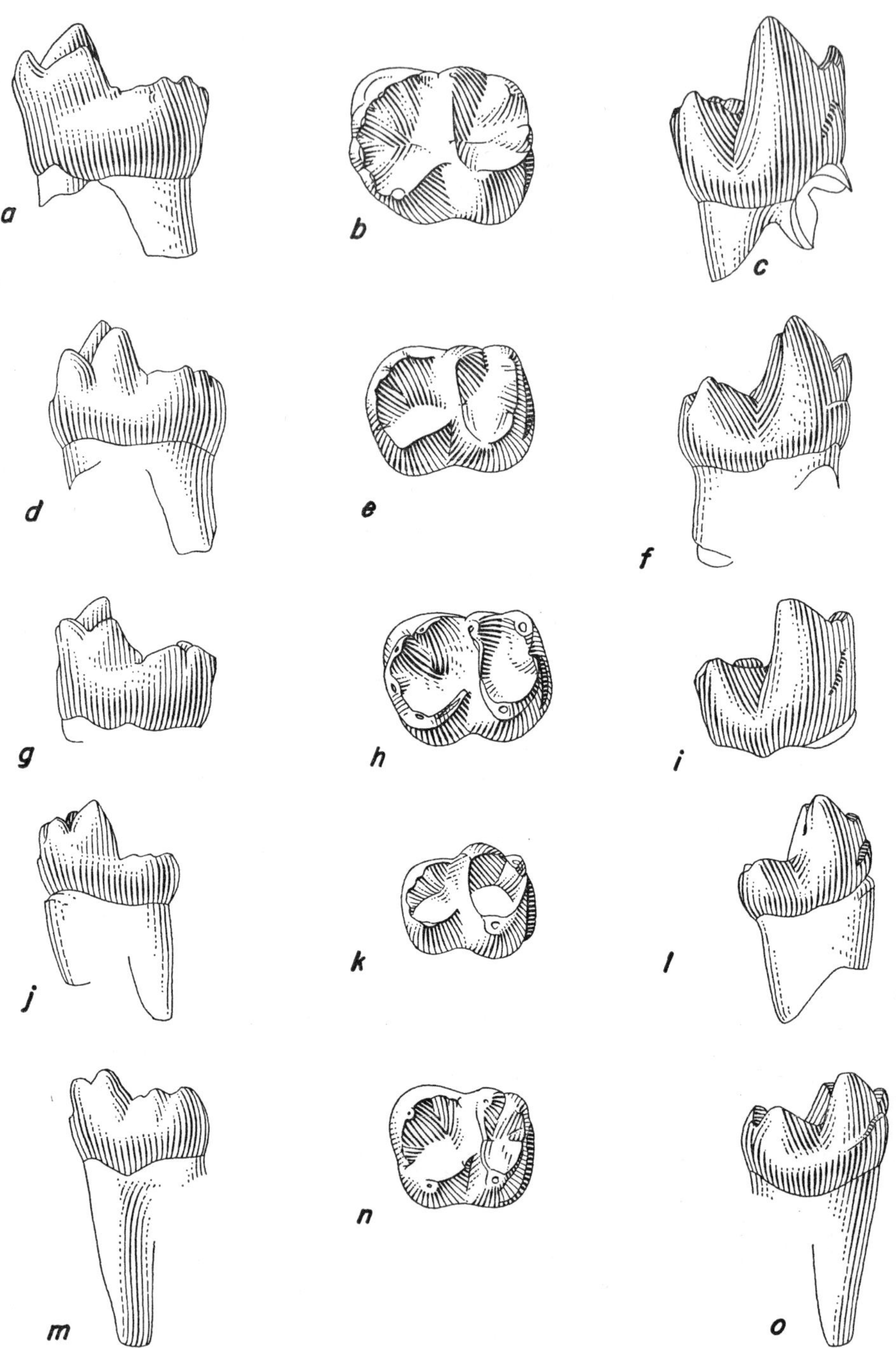

Fig. 17. cf. Paroxyclaenidae, X 6. Right M₁?, M.N.H.N.-Av 6759, *a*, lingual view, *b*, occlusal view, and *c*, labial view. Right M₂?, M.N.H.N.-Louis-189 Mu, *d*, lingual view, *e*, occlusal view, *f*, labial view. Right M₂?, M.N.H.N.-Av 5731, *g*, lingual view, *h*, occlusal view, and *i*, labial view. Right M₃?, M.N.H.N.-Louis-49 **Py**, *j*, lingual view, *k*, occlusal view, and *l*, labial view. Right M₃?, M.N.H.N.-190 Mu, *m*, lingual view, *n*, occlusal view, and *o*, labial view.

varies from slightly to markedly lower and from slightly larger to much smaller than the metaconid. The paraconid though not twinned with the metaconid is in close proximity and is located anterior and somewhat labial to it. The protoconid is usually lower but may be equal in height to the metaconid, becoming markedly higher on teeth in advanced stages of wear; its labial height above the talonid varies from three-fourths to equal the trigonid length. The protoconid is equal to somewhat smaller than the metaconid in basal dimensions. It is located one-fourth the trigonid width lingual to the labial border of the tooth and is labial to antero-labial to the metaconid. The connate bases of the paraconid and metaconid form the lingual wall of the prefossid, a distinct paralophid forms the anterior wall, and a weak protolophid leaves the posterior side open. A small carnassial notch is developed on the protolophid of one specimen about midway between the para-conid and the metaconid. The anterior cingulum extends from a point on the prevallid below the paraconid to another point on the same surface ventral and slightly lateral to the protoconid. The anterior cingulum dips 15 to 60 degrees below the horizontal.

The talonid is short anteroposteriorly, the length being one-half to two-thirds the width. The talonid is subequal to the trigonid in width and length. The ento-cristid and postcristid form a posterolingually convex ridge from the base of the metaconid to the hypoconid. The crista obliqua is either straight or somewhat medially concave and directed towards the trigonid, abutting its base at a point one-half to two-thirds the distance from the metaconid to the protoconid. The hypoconid is always distinct and although the lowest talonid cusp, it is the largest in basal dimensions. Its position may be anywhere from near the lateral border of the tooth to one-fourth the talonid width lingual to the labial edge. The other talonid cusps are mere swellings on the postcristid and entocristid. When present, the hypoconulid appears as a small swelling on the posterior border of the tooth near the midline. The entoconid is situated on the posterior border of the talonid. Located medial to the hypoconid near the lingual edge of the tooth, the ento-conulid is smaller than and anteromedial to the entoconid. The postfossid is well enclosed and equal to the prefossid in depth. Between the protoconid and hypo-conid there is no trace of an ectostylid with the exception of one specimen (M.N.H.N.–Mu 6168) where the cusp is well developed. No distinct wear facets are present on the vertical surfaces of the tooth, all the abraded areas being horizontal planes at the apices of the trigonid cusps.

COMMENTS

The diversity in size and morphology of these specimens suggests that more than one lower molar position and (or) species is represented in this assemblage.

The M_2 of *Spaniella carezi* Crusafont and Russell, 1967 (see Crusafont and Russell, 1967, figs. 2, 3) is similar to these lower molars in the anteroposterior shortening of the trigonid; position, size, and relative heights of the trigonid cusps; strength, position, and orientation of the anterior cingulum; development of the protolophid and paralophid; similarity in the placement of the unbroken cristids that completely enclose the postfossid; and relative heights of the hypo-conid and entoconid.

In contrast to these lower molars, the M_2 of *Spaniella carezi* has a slightly longer talonid relative to the length of the trigonid, the trigonid is lower and its posterointernal edge is inclined at an angle of about 45 degrees to the base of the enamel rather than being nearly vertical. Although these differences exclude these lower molars from assignment to *S. carezi*, they do not exclude them from the Paroxyclaenidae for other species within that family also differ from *S. carezi* in the same ways (cf. *Kopidodon macrognathus* and *Kochictis centenii*).

The specimens identified above as $M_{3?}$ are smaller than those identified as $M_{2?}$ but resemble more closely the M_2 of *Spaniella carezi* rather than its M_3. If the identification of the position in the tooth row of those specimens designated $M_{1?}$ is correct, it is possible that the trigonid of the unknown M_1 of *S. carezi* is no more anteroposteriorly expanded than the trigonid of the M_2.

These specimens are similar to the M_2 of individuals referred to *Vulpavus australis* from the Wasatchian of New Mexico (A.M.N.H. 16226, see Matthew, 1915*d*, fig. 33) and Wyoming (A.M.N.H. 2369 and 2813). They also resemble the M_2 of several Wasatchian specimens variously identified as *V.* cf. *V. australis* and *Vulpavus* from Wyoming (A.M.N.H. 5024, 2086, and 2087) and from New Mexico (A.M.N.H. 55420). The resemblance exists in the similar proportions of trigonid length and width; the close proximity and relative location of the paraconid and metaconid to one another; the smaller basal dimensions of the paraconid with respect to the metaconid; the presence and development of an anterior cingulum; the development and arrangement of talonid cristids; and the development of numerous talonid cusps, not just a strong hypoconid.

The North American specimens of *Vulpavis australis* M_2 display a carnassial notch in the paralophid, a feature present on only one of the French specimens. Another more disturbing fact is the absence of a *Vulpavus* M_1 in the identified assemblage of specimens. It seems improbable that fourteen specimens of the M_2 would be found without a single example of the larger M_1 being discovered.

Order incertae sedis
Family incertae sedis
cf. *Hyracolestes* Matthew and Granger, 1925*a*

TABLE 32

MEASUREMENTS OF CF. HYRACOLESTES M_1

Specimens	Locality	Trigonid width	Trigonid length	Talonid width	Total length
M.N.H.N.-Mu 6601 Left M_1 trigonid.....	Mutigny	2.7	2.4	...	...
M.N.H.N.-Av 4614 Left M_1...............	Avenay	3.0	2.9	2.5	5.1

DESCRIPTION

(Fig. 18*a–c*)

The length of the trigonid is from four-fifths to subequal the width. The paraconid is lower than the metaconid but equal in basal dimensions and is located anterior and somewhat lateral to it. The base of the paraconid projects anteriorly from the mesial root of the tooth. The protoconid is the highest and largest trigonid cusp; its labial height above the talonid is about equal to the trigonid length. It is located at the labial edge of the tooth lateral and slightly anterior to the metaconid. The tips of the trigonid cusps are truncated by wear and one cannot determine whether they were sharp or rounded in the unworn condition. The paralophid is more prominent and has a larger carnassiform notch than does the protolophid. The prefossid is deep and open lingually with a prominent separation between the paraconid and metaconid. The anterior cingulum dips downward only slightly in the lateral direction and extends on the prevallid from its lingual edge to a point just below the protoconid. A slight appression fossette is developed at the lingual end of the anterior cingulum.

The talonid is wider than long, five-sixth as wide as the trigonid and three-fourths as long. The weak entocristid is only a slight swelling on the lingual border of the talonid. It begins about one-third of the talonid length behind the base of the metaconid and extends posterolaterally meeting the postcristid which continues to the hypoconulid where it abruptly turns anterolabially extending to

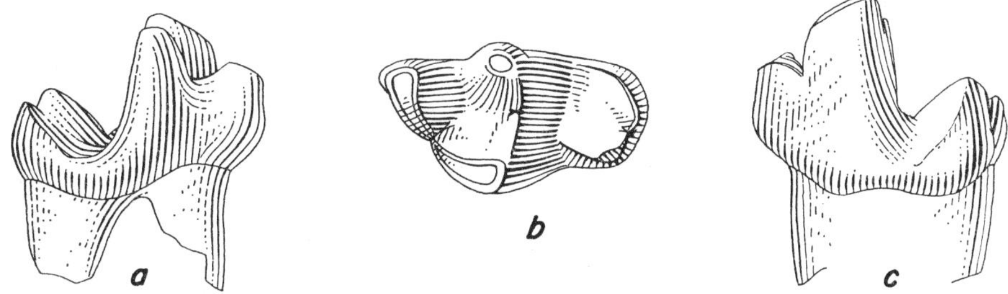

Fig. 18. cf. *Hyracolestes*, X 6, left M₁, M.N.H.N-Av 4614, *a*, lingual view, *b*, occlusal view, and *c*, labial view.

the hypoconid. The crista obliqua is directed straight towards the trigonid from the hypoconid but does not quite contact it; were this crista obliqua extended, it would abut the trigonid base at a point medial to the protoconid not more than one-fourth the distance to the metaconid. The hypoconid is the highest and largest talonid cusp, located on the lateral border of the tooth and separated from the hypoconulid by a deep notch in the postcristid. The hypoconulid is low and small, a mere swelling of the postcristid near the middle of the posterior edge of the tooth. There is only the faintest indication of an entoconid and no entoconulid. The postfossid is deep; it is open widely on the lingual side and less so on the labial side. The surface of the postfossid is an extension of the steeply inclined anteromedial side of the hypoconid. This plane continues without change of direction or inclination to the lingual border of the talonid. There is no trace of an ectostylid. Prominent wear facets are developed on the dorsal half of the prevallid, on the entire postvallid and on the anterolateral face of the hypoconid.

COMMENTS

These teeth are closely comparable to the most posteriorly preserved molar in the type and only specimen of *Hyracolestes erminenus* Matthew and Granger,

1925*a* (A.M.N.H. 20425, see Matthew and Granger, 1925*a*, fig. 13)[1] from the (?) early Eocene of Mongolia. The only readily observed difference is the much greater size of the French specimens. Every measurement on *H. ermineus* is approximately doubled on these specimens so that the proportions remain the same. The details of trigonid and talonid morphology are strikingly similar so that to enumerate most of them would be to repeat the description given above. There are three slight morphological differences however. *H. ermineus* lacks an anterior cingulum at the base of the prevallid. On the complete French specimen the hypoconulid is on the midline and not well separated from the hypoconid, whereas this cusp is slightly lingual to the midline and well separated from the hypoconid on *H. ermineus.*

The weakness of the entocristid together with the steeply inclined postfossid surface of the specimen from France indicates a pattern of occlusion different from that found in the proviverrines *Prototomus* and *Proviverra.*

The most noticeable differences between these teeth and those of miacids are to be found in the construction of the paraconid. The paraconid is much lower than the metaconid, and the base of the cusp projects anteriorly from the mesial root of the tooth. These characters are developed on some miacids but not to the degree found here.

CONCLUSIONS

At least three different miacid genera are known from the Sparnacian of France: cf. *Viverravus,* cf. *Miacis,* and cf. *Uintacyon.* All three occur in North America and, if correctly identified, indicate interchange between the two areas. However, the presence of several specimens which could not be identified below the familial level suggests the existence of an endemic element among the French miacids. To typify adequately a new miacid genus requires knowledge of almost complete dentitions, not merely isolated teeth. Inasmuch as my identifications were based solely on isolated teeth, tentative generic assignments were possible only in cases where comparisons could be made with adequate material. At present, most of the adequate material known was collected in North America. Therefore, the similarity of the miacid faunas of Europe and North America cannot be meaningfully measured with the specimens currently available. The absence of the small-tooth miacids from the Cuisian sites is probably the result of collecting bias because little underwater sieving has been carried out at any of the three localities.

Three deltatheridian genera are recognized on sufficiently good material that they may be used to estimate faunal similarities between France and other regions: *Prototomus s. l., Francotherium* n. gen., and *Oxyaena. Prototomus s. l.* is known from the Sparnacian and Cuisian. *Oxyaena* also occurs in both stages if the Sinceny locality is Sparnacian. Both *Prototomus s. l.* and *Oxyaena* are well known in the North American Eocene but *Francotherium* has never been found in the western hemisphere. This might be the result of the later appearance of *Francotherium* in the Cuisian after the corridor permitting extensive faunal

---

[1] This specimen is being refigured by Szalay and McKenna in their forthcoming review of the primitive mammals from Gashato.

interchange between North America and Europe during the Sparnacian was restricted or broken. Another deltatheridian genus, cf. *Didelphodus,* can be tentatively identified in the assemblage of Sparnacian mammals from France, but the material is insufficient for the measurement of faunal similarity.

Phenacodus is well documented from both Sparnacian and Cuisian deposits and is widely known from the North American late Paleocene and early Eocene. Cf. *Paroxyclaenus* and another, possibly new genus of the Eurasian family Paroxyclaenidae are present in the Sparnacian of France.

Hyracolestes, an enigmatic genus of uncertain familial and ordinal assignment, from the Mongolian Gashato local faunule of (?) early Eocene age may have a relative in the Sparnacian of Europe. However, the scanty nature of the material makes the assignment tentative at best.

In summary, the assemblage of specimens described in this report reflects the dual nature of the early Eocene mammal fauna of Europe. Seven of the genera discussed above are well known in North America[2] while the remaining four are exclusively European or Old World forms.[3] Because of the fragmentary nature of the majority of this material, only two specimens, the jaws of *Oxyaena menui* and *Phenacodus* cf. *P. teilhardi,* can be used as meaningful evidence for the problem of the correlation of the European and North American mammal faunas. Important similarities are present between the jaw of the European Sparnacian *O. menui* and those of the North American Wasatchian *Oxyaena gulo.* However, because the lineages which gave rise to the two species appear to have diverged by late Paleocene time, the correlation of the Cuisian with some part of the Wasatchian is only weakly supported by this evidence. In contrast, the similarities between the specimens of the North American Wasatchian *Phenacodus intermedius* and the jaw here referred to *P.* cf. *P. teilhardi* are so great that the two species can be thought of as closely related if not the same. This high similarity supports the correlation of the European Sparnacian with some part of the North American Wasatchian. None of the other specimens described in this report, if identified correctly, would tend to contradict these correlations.

---

[2] Cf. *Didelphodus, Prototomus s. l., Oxyaena,* cf. *Viverravus,* cf. *Miacis,* cf. *Uintacyon,* and *Phenacodus.*

[3] *Francotherium,* cf. *Paroxyclaenus,* unnamed genus within the Paroxyclaenidae, and cf. *Hyracolestes.*

REFERENCES CITED

The alphabetic designations given to papers published prior to 1964 are those used in the bibliographies of fossil vertebrates, Hay 1902a and 1930a, Romer et al., 1962, and Camp et al., 1940, 1942, 1949, 1953, 1961, 1964, and 1968.

BLUMENBACH, J. F.

1791. Handbuch de Naturgeschichte. Vierte auflage. Göttingen: Johann Christian Dieterich. xii + 704 + (33) pp., 3 pls.

BOWDICH, T. E.

1821. An analysis of the natural classification of *Mammalia,* for the use of students and travellers. Paris: J. Smith. 115 + (31) pp., 15 pls.

CAMP, C. L., and H. J. ALLISON

1961. Bibliography of fossil vertebrates, 1949–1953. Geol. Soc. Amer., Mem. no. 84. 532 pp.

CAMP, C. L., H. J. ALLISON, and R. H. NICHOLS

1964. Bibliography of fossil vertebrates, 1954–1958. Geol. Soc. Amer., Mem. no. 92. 647 pp.

CAMP, C. L., H. J. ALLISON, R. H. NICHOLS, and H. McGINNIS

1968. Bibliography of fossil vertebrates, 1959–1963. Geol. Soc. Amer., Mem. no 117. 644 pp.

CAMP, C. L., D. N. TAYLOR, and S. P. WELLES

1942. Bibliography of fossil vertebrates, 1934–1938. Geol. Soc. Amer., Sp. Pap., no. 42. 663 pp.

CAMP, C. L., and V. L. VANDERHOOF

1940. Bibliography of fossil vertebrates, 1928–1933. Geol. Soc. Amer., Sp. Pap., no. 27. 503 pp.

CAMP, C. L., S. P. WELLES, and MORTON GREEN

1949. Bibliography of fossil vertebrates, 1939–1943. Geol. Soc. Amer., Mem. no. 37. 371 pp.

1953. Bibliography of fossil vertebrates, 1944–1948. Geol. Soc. Amer., Mem. no. 57. 465 pp.

COOPER, C. F.

1932b. On some mammalian remains from the Lower Eocene of the London Clay. Ann. Mag. Nat. Hist., tenth series, 9:458–467, 2 figs., 2 pls., tables.

COPE, E. D.

1872e. Third account of new *Vertebrata* from the Bridger Eocene of Wyoming Territory. Proc. Amer. Philos. Soc., 12:469–472 (1873).

1873d. Fourth notice of extinct *Vertebrata* from the Bridger and the Green River Tertiaries. Palaeont. Bull. No. 17:1–4.

1874o. Report upon vertebrate fossils discovered in New Mexico, with descriptions of new species. Extract from Appendix FF of Annual Report of the Chief of Engineers, 1874. Washington: Gov. Print. Office, pp. 1–18.

1877k. Report upon the extinct *Vertebrata* obtained in New Mexico by parties of the expedition of 1874. Chapter XI; Fossils of the Mesozoic periods and Geology of Mesozoic and Tertiary beds. XII; Fossils of the Eocene period. XIII; Fossils of the Loup Fork Epoch. Geogr. Surv. west of 100th meridian, Corps of Engineers, U. S. Army, vol. 4, part 2, Paleontology, pp. 1–370, pls. 22–83.

1880c. On the genera of the *Creodonta.* Proc. Amer. Philos. Soc., 19:76–82.

1881x. A new type of Perissodactyla. Amer. Naturalist, 15:1017–1018.

1882e. Contributions to the history of the *Vertebrata* of the Lower Eocene of Wyoming and New Mexico, made during 1881. I. The fauna of the Wasatch beds of the basin of the Big Horn River. II. The fauna of the *Catathlaeus* beds, or Lowest Eocene, New Mexico. Proc. Amer. Philos. Soc., 20:139–197.

1882x. Notes on Eocene *Mammalia.* Amer. Naturalist, 16:552.

1884o. The *Vertebrata* of the Tertiary formations of the West. Book I. Report U. S. Geolog. Survey of the Territories, F. V. Hayden, U. S. geologist in charge, 1884, Washington, 3:i-xxxv, 1–1009; pls. 1–75a. This work actually issued in 1885.

CRUSAFONT PAIRÓ, M.
  1956b. Otro nuevo Condilartro del Luteciense Pirenaico. Boll. Soc. Geol. Ital., 75:42–47, 1 fig.
  1957c. Adición a los mamiferos fósiles del Luteciense de Montllobar (Tremp). Cursillos y Conf. Inst. "Lucas Mallada," no. 4:71–73.
CRUSAFONT PAIRÓ, M., and D. E. RUSSELL
  1967. Un nouveau paroxyclaenidé de l'Eocene d'Espagne. Bull. Mus. Hist. Nat., Paris, second ser., 39(4):757–773, figs. 1–4.
CRUSAFONT PAIRÓ, M., and J. F. de VILLALTA COMELLA
  1954b. "*Almograver*," un nuevo primate del eoceno pirenaico. Est. Geol. Inst. Invest., Lucas Mallada, 10 (22):165–176, pls. 13–14.
DEHM, R. and T. ZU OETTINGEN-SPIELBERG
  1958. Paläontologische und geologische Untersuchungen im Tertiär von Pakistan. 2. Die mitteleocänen Säugetiere von Ganda Kas bei Basal in Nordwest-Pakistan. Abhandl. Bayerischen Akad. Wiss. (Math.-Naturw. Kl., N.F.), 91:1–54.
DENISON, R. H.
  1938. The broad-skulled Pseudocreodi. Ann. N. Y. Acad. Sci., 37:163–256, 32 figs.
DORR, J. A.
  1958a. Early Cenozoic vertebrate paleontology, sedimentation, and orogeny in central western Wyoming. Bull. Geol. Soc. Amer., 69:1217–1243, 4 figs., 2 pls., 4 tables.
DOUGLASS, E.
  1901b. Fossil Mammalia of the White River beds of Montana. Trans. Amer. Philos. Soc., new series, 20:237–279, 1 map, pl. 60.
GAZIN, C. L.
  1962. A further study of the lower Eocene mammalian faunas of southwestern Wyoming. Smithson, Misc. Coll., vol. 144, no. 1, v + 98 pp., 2 figs., 14 pls.
GILL, T.
  1872b. Arrangement of the families of mammals and synoptical tables of characters of the subdivisions of mammals. Smithson, Misc. Coll., 11 (art. 1):i–vi, 1–98.
GRANGER, W.
  1915a. A revision of the Lower Eocene Wasatch and Wind River faunas. Part III; Order Condylarthra. Families Phenacondontidae and Meniscotheriidae. Bull. Amer. Mus. Nat. Hist., 34:329–361, 18 text-figs.
HAY, O. P.
  1902a. Bibliography and catalogue of the fossil vertebrata of North America. Bull. U. S. Geol. Surv., no. 179:1–868.
  1930a. Second bibliography and catalogue of the fossil vertebrata of North America. Carnegie Inst. Wash. Publ. 390. Vol. 1 (1929), vii, 916 pp.; vol. 2 (1930), xiv, 1074 pp.
HIBBARD, C. W.
  1949d. Techniques of collecting microvertebrate fossils. Contrib. Mus. Pal. Univ. Mich., 8:7–19, 4 pls.
KRETZOI, M.
  1943a. *Kochictis contenii* n. g., n. sp. az egeresi felsö Oligocenböl. Föld. Közlöny, 73:10–17, 1 pl.
         Also in German: *Kochictis centenii* n. g., n. sp., ein altertümlicher Creodonte aus dem Oberoligozän Siebenbürgens. Ibid., pp. 190–195.
LEIDY, J.
  1869a. The extinct mammalian fauna of Dakota and Nebraska, including an account of some allied forms from other localities, together with a synopsis of the mammalian remains of North America. Jour. Acad. Nat. Sci. Phila., second series, 7:1–472, with 30 plates.
  1872j. Remarks on fossils from Wyoming. Proc. Acad. Nat. Sci. Phila., 1872, p. 277.
LEMOINE, V.
  1880a. Sur les ossements fossiles des terrains tertiaires inférieurs des environs de Reims. Assoc. Française pour l'avancement des Sciences, 8e session, Montpellier, 1879, pp. 585–594.
LINNAEUS, C.
  1758a. Systema naturae. 10th ed. Stockholm, vol. 1. 824 pp.

MacIntyre, G. T.

  1966. The Miacidae (Mammalia, Carnivora). Part 1. The Systematics of *Ictidopappus* and *Protictis*. Bull. Amer. Mus. Nat. Hist., 131 (art. 2):115–210, figs. 1–21, pls. 1–20, tables 1–12.

McKenna, M. C.

  1960*h*. Fossil Mammalia from the Early Wasatchian Four Mile fauna, Eocene of northwest Colorado. Univ. Calif. Publ. Geol. Sci., 37(1):1–130, figs. 1–64.

  1962*b*. Collecting small fossils by washing and screening. Curator, 5(3):221–235, 6 figs.

Marsh, O. C.

  1872*g*. Preliminary description of new Tertiary mammals. Pt. I. Amer. Jour. Sci., third series, 4:122–128 and erratum on p. 504.

Matthes, H. W.

  1952. Die Creodontier aus der mitteleozänen Braunkohle des Geiseltales. Hallesches. Jahrb. Mitteldeutsche Erdgeschichte, 1:201–240, 4 figs.

Matthew, W. D.

  1909*d*. The Carnivora and Insectivora of the Bridger Basin, Middle Eocene. Mem. Amer. Mus. Nat. Hist., 9 (part 6):291–567, pls. 43–52, figs. 1–118.

  1915*d*. A revision of the Lower Eocene Wasatch and Wind River faunas. Part I: Order Ferae (Carnivora). Suborder Creodonta. Bull. Amer. Mus. Nat. Hist., 34:1–103, figs. 1–87.

  1915*f*. A revision of the Lower Eocene Wasatch and Wind River faunas. Part IV: Entelonychia, Primates, Insectivora (part). Bull. Amer. Mus. Nat. Hist., 34:429–483, pl. 15, figs. 1–52.

  1918*h*. A revision of the Lower Eocene Wasatch and Wind River faunas. Insectivora (continued), Glires, Edentata. Bull. Amer. Mus. Nat. Hist. 38:565–657, figs. 1–68.

Matthew, W. D. and W. Granger

  1925*a*. Fauna and correlation of the Gashato formation of Mongolia. Amer. Mus. Novit., no. 189:1–12, figs. 1–14.

  1925*d*. New mammals from the Irdin Manha Eocene of Mongolia. Amer. Mus. Novit., no. 198:1–10, figs. 1–10.

Osborn, H. F.

  1910*b*. The age of mammals in Europe, Asia and North America. New York: Macmillan, i–xvii + 1–635 pages, figs. 1–220.

Parker, T. J., and W. A. Haswell

  1897*a*. A text-book of zoology. London: Macmillan. Vol. 2, xx + 683 pp., figs. 664–1172.

Robinson, P.

  1966. Fossil Mammalia of the Huerfano Formation, Eocene, of Colorado. Bull. Peabody Mus. Nat. Hist., Yale Univ., no. 21:1–95, figs. 1–9, pls. 1–10.

Romer, A. S., N. E. Wright, T. Edinger, and Richard Van Frank

  1962. Bibliography of fossil vertebrates exclusive of North America, 1509–1927. Geol. Soc. Amer., Mem. 87, 2 vols., lxxxix + 1544 pp.

Russell, D. E., P. Louis, and D. E. Savage

  1967. Primates of the French Early Eocene. Univ. Calif. Publ. Geol. Sci., 73:1–46, figs. 1–14.

Rütimeyer, L.

  1888*a*. Ueber einzige Beziehungen zwischen den Säugetheirstämmen alter und neuer Welt. Erster Nachtrag zu der eocänen Fauna von Egerkingen. Abh. schweiz. pal. Ges., 15: 1–63, pl. 1.

Savage, D. E., D. E. Russell, and P. Louis

  1965. European Eocene Equidae. Univ. Calif. Publ. Geol. Sci., 56:1–94, figs. 1–42, pl. 1.

  1966. Ceratomorpha and Ancylopoda (Perissodactyla) from the Lower Eocene Paris Basin, France. Univ Calif. Publ. Geol. Sci., 66:1–38, figs. 1–28.

Simpson, G. G.

  1929*g*. Paleocene and Lower Eocene mammals of Europe. Amer. Mus. Novit., no. 354:1–17.

  1931*b*. A new classification of mammals. Bull. Amer. Mus. Nat. Hist., 59:259–293.

  1935*g*. New Paleocene mammals from the Fort Union of Montana. Proc. U. S. Nat. Mus., 83: 221–244.

1937*d*. Note on the Clark Fork, Upper Paleocene, fauna. Amer. Mus. Novit., no. 954:1–24, figs. 1–6.

1945*i*. The principles of classification and a classification of mammals. Bull. Amer. Mus. Nat. Hist., 85, v-xvi + 350 pp.

1947*a*. Note on the measurement of variability and on relative variability of teeth of fossil mammals. Amer. Jour. Sci., 245:522–525.

TEILHARD DE CHARDIN, P.

1921*a*. Les mammifères de l'Éocène inférieur français et leurs gisements. Ann. Pal. (Paris), X:171–176, 2 figs. (See 1922*a*.)

1922*a*. Les mammifères de l'Éocène inférieur français et leurs gisements. Ann. Pal. (Paris), XI:9–116, 40 figs., 8 pls. (See 1921*a*.)

1927*a*. Les mammifères de l'Éocène inférieur de la Belgique. Mém. Mus. Hist. nat. Belgique, no. 36:1–33, figs. 1–29, pls. 1–6.

TROUESSART, É.

1885*a*. Catalogue des mammifères vivants et fosslies. (Carnivora.) Bull. Soc. Études sci. Angers, 14 (Suppl.):1–108.

VAN VALEN, L.

1965*a*. Paroxyclaenidae, an extinct family of Eurasian mammals. Jour. Mammal., 46(3):388–397, figs. 1–4, table 1.

1965*b*. Some European Proviverrini (Mammalia, Deltatheridia), Pal., 8 (pt. 4):638–665, figs. 1–6, tables 1–2.

1966. Deltatheridia, a new order of mammals. Bull. Amer. Mus. Nat. Hist., 132:1–126, figs. 1–17, pls. 1–8, tables 1–26.

1967. New Paleocene insectivores and insectivore classification. Bull. Amer. Mus. Nat. Hist., 135:217–284, figs. 1–7, pls. 6–7, tables 1–7.

WEITZEL, K.

1933*d*. *Kopidodon macrognathus* Wittich, ein Raubtier aus dem Mitteleozän von Messel. Notizbl. Ver. Erdk., fifth series, 14:81–88, 1 pl.

WITTICH, E.

1902*a*. *Cryptopithecus macrognathus* n. spec., ein neuer Primate aus den Braunkohlen von Messel. Centralbl. Min. Geol. Pal., pp. 289–294, figs. 1–3.

WORTMAN, J. L.

1902*a*. Studies of Eocene Mammalia in the Marsh collection, Peabody Museum. Amer. Jour. Sci., fourth series, 13:39–46, 115–128, 197–206, 433–448, figs. 61–95, pls. 9–10; 14:17–23, figs. 96–99.